LINEAR ALGEBRA

Johns Hopkins Studies in the
Mathematical Sciences

in association wth the
Department of Mathematical Sciences
The Johns Hopkins University

LINEAR ALGEBRA

Challenging Problems for Students

Second Edition

Fuzhen Zhang

The Johns Hopkins University Press
Baltimore

Printed in the United States of America on acid-free paper
9 8 7 6 5 4 3 2 1

The Johns Hopkins University Press
2715 North Charles Street
Baltimore, Maryland 21218-4363
www.press.jhu.edu

Library of Congress Control Number: 2008936105

A catalog record for this book is available from the British Library.

The Johns Hopkins University Press uses environmentally friendly book materials, including recycled text paper that is composed of at least 30 percent post-consumer waste, whenever possible. All of our book papers are acid-free, and our jackets and covers are printed on paper with recycled content.

To the memory of my grandfather and parents

Contents

Preface to the 2nd Edition

This is the second, revised, and expanded edition of the linear algebra problem book *Linear Algebra: Challenging Problems for Students.* The first edition of the book, containing 200 problems, was published in 1996. In addition to about 200 new problems in this edition, each chapter starts with definitions and facts that lay out the foundations and groundwork for the chapter, followed by carefully selected problems. Some of the new problems are straightforward; some are pretty hard. The main theorems frequently needed for solving these problems are listed on page xv.

My goal has remained the same as in the first edition: to provide a book of interesting and challenging problems on linear algebra and matrix theory for upper-division undergraduates and graduate students in mathematics, statistics, engineering, and related fields. Through working and practicing on the problems in the book, students can learn and master the basic concepts, skills, and techniques in linear algebra and matrix theory.

During the past ten years or so, I served as a collaborating editor for *American Mathematical Monthly* problem section, associate editor for the *International Linear Algebra Society Bulletin* IMAGE Problem Corner, and editor for several other mathematical journals, from which some problems in the new edition have originated. I have also benefited from the math conferences I regularly attend; they are the International Linear Algebra Society (ILAS) Conferences, Workshops on Numerical Ranges and Numerical Radii, R. C. Thompson (formerly Southern California) Matrix Meetings, and the International Workshops on Matrix Analysis and Applications. For example, I learned Problem 4.21 from M.-D. Choi at the ILAS Shanghai Meeting in 2007; Problem 4.97 was a recent submission to IMAGE by G. Goodman and R. Horn; some problems were collected during tea breaks.

I am indebted to many colleagues and friends who helped with the revision; in particular, I thank Jane Day for her numerous comments and suggestions on this version. I also thank Nova Southeastern University (NSU) and the Farquhar College of Arts and Sciences (FCAS) of the university for their support through various funds, including the President's Faculty Research and Development Grants (Awards), FCAS Minigrants, and FCAS Faculty Development Funds.

Readers are welcome to communicate with me at zhang@nova.edu.

Preface

This book is written as a supplement for undergraduate and first-year graduate students majoring in mathematics, statistics, or related areas. I hope that the book will be helpful for instructors teaching linear algebra and matrix theory as well.

Working problems is a crucial part of learning mathematics. The purpose of this book is to provide a suitable number of problems of appropriate difficulty. The readers should find the collection of two hundred problems in this book diverse, interesting, and challenging.

This book is based on my ten years of teaching and doing research in linear algebra. Although the problems have not been systematically arranged, I have tried to follow the order and level of some commonly used linear algebra textbooks. The theorems that are well known and found in most books are excluded and are supposed to be used freely. The problems vary in difficulty; some of them may even baffle professional experts. Only a few problems need the Jordan canonical forms in their solutions. If you have a little elementary linear algebra background, or are taking a linear algebra course, you may just choose a problem from the book and try to solve it by any method. It is expected that readers will refer to the solutions as little as possible.

I wish to dedicate the book to the memory of my Ph.D. advisor, R. C. Thompson, a great mathematician and a founder of the International Linear Algebra Society (ILAS). I am grateful to C. A. Akemann, R. A. Horn, G. P. H. Styan, B.-Y. Wang, and X.-R. Yin for guiding me toward the road of a mathematician. I would also like to thank my colleagues J. Bartolomeo, M. He, and D. Simon for their encouragement. Finally, I want to thank Dr. R. M. Harington, of the Johns Hopkins University Press, for his enthusiastic cooperation.

Frequently Used Notation and Terminology

$\mathbb{R}$	real number field
$\mathbb{C}$	complex number field
$\mathbb{F}$	scalar field $\mathbb{R}$ or $\mathbb{C}$
$\mathbb{R}^n$	vectors of n real components
$\mathbb{C}^n$	vectors of n complex components
$M_n(\mathbb{F})$	$n \times n$ matrices with entries from $\mathbb{F}$
$M_{m\times n}(\mathbb{F})$	$m \times n$ matrices with entries from $\mathbb{F}$
$\dim V$	dimension of vector space V
I	identity matrix
$A = (a_{ij})$	matrix A with entries a_{ij}
$r(A)$	rank of matrix A
$\operatorname{tr} A$	trace of matrix A
$\det A$	determinant of matrix A
$\lvert A \rvert$	determinant of matrix A (particularly for block matrices)
A^{-1}	inverse of matrix A
A^t	transpose of matrix A
$\bar{A}$	conjugate of matrix A
A^*	conjugate transpose of matrix A, i.e., $A^* = \bar{A}^t$
$\operatorname{Ker} A$	kernel or null space of A, i.e., $\operatorname{Ker} A = \{\, x \mid Ax = 0 \,\}$
$\operatorname{Im} A$	image or range of A, i.e., $\operatorname{Im} A = \{Ax\}$
$A \geq 0$	A is positive semidefinite
$A \geq B$	$A - B$ is positive semidefinite
$\operatorname{diag}(\lambda_1, \lambda_2, \ldots, \lambda_n)$	diagonal matrix with $\lambda_1, \lambda_2, \ldots, \lambda_n$ on the main diagonal
$A \circ B$	Hadamard product of matrices A and B, i.e., $A \circ B = (a_{ij}b_{ij})$
$\langle u, v \rangle$	inner product of vectors u and v
$\lVert x \rVert$	norm or length of vector x

An $n \times n$ matrix A is said to be

upper-triangular	if all entries below the main diagonal are zero
diagonalizable	if $P^{-1}AP$ is diagonal for some invertible matrix P
similar to B	if $P^{-1}AP = B$ for some invertible matrix P
unitarily similar to B	if $U^*AU = B$ for some unitary matrix U
unitary	if $AA^* = A^*A = I$
positive semidefinite	if $x^*Ax \geq 0$ for all vectors $x \in \mathbb{C}^n$
Hermitian	if $A = A^*$
normal	if $A^*A = AA^*$, and

a scalar λ is an eigenvalue of A if $Ax = \lambda x$ for some nonzero vector x

Frequently Used Theorems

- **Dimension identity:** Let W_1 and W_2 be subspaces of a finite dimensional vector space V. Then

$$\dim W_1 + \dim W_2 = \dim(W_1 + W_2) + \dim(W_1 \cap W_2).$$

- **Theorem on the eigenvalues of AB and BA:** Let A and B be $m \times n$ and $n \times m$ complex matrices, respectively. Then AB and BA have the same nonzero eigenvalues, counting multiplicity. Thus

$$\operatorname{tr}(AB) = \operatorname{tr}(BA).$$

- **Schur triangularization theorem:** For any square matrix A, there exists a unitary matrix U such that U^*AU is upper-triangular.

- **Jordan decomposition theorem:** Let A be an $n \times n$ complex matrix. Then there exists an $n \times n$ invertible matrix P such that

$$A = P^{-1} \operatorname{diag}(J_1, J_2, \dots, J_k)P,$$

where each J_i, $i = 1, 2, \dots, k$, is a Jordan block.

- **Spectral decomposition theorem:** Let A be an $n \times n$ normal matrix with eigenvalues $\lambda_1, \lambda_2, \dots, \lambda_n$. Then there exists an $n \times n$ unitary matrix U such that

$$A = U^* \operatorname{diag}(\lambda_1, \lambda_2, \dots, \lambda_n)U.$$

In particular, if A is positive semidefinite, then all $\lambda_i \geq 0$; if A is Hermitian, then all λ_i are real; and if A is unitary, then all $|\lambda_i| = 1$.

- **Singular value decomposition theorem:** Let A be an $m \times n$ complex matrix with rank r. Then there exist an $m \times m$ unitary matrix U and an $n \times n$ unitary matrix V such that

$$A = UDV,$$

where D is the $m \times n$ matrix with (i, i)-entries the singular values of A, $i = 1, 2, \dots, r$, and other entries 0. If $m = n$, then D is diagonal.

- **Cauchy-Schwarz inequality:** Let V be an inner product space over a number field ($\mathbb{R}$ or $\mathbb{C}$). Then for all vectors x and y in V

$$|\langle x, y\rangle|^2 \leq \langle x, x\rangle \langle y, y\rangle.$$

Equality holds if and only if x and y are linearly dependent.

Linear Algebra

Chapter 1

Vector Spaces

Definitions and Facts

Vector Space. A vector space involves four things – two (nonempty) sets V and $\mathbb{F}$ and two algebraic operations called vector addition and scalar multiplication. The objects in V are called *vectors* and the elements in $\mathbb{F}$ are *scalars*. In this book, $\mathbb{F}$ is either the field $\mathbb{R}$ of real numbers or the field $\mathbb{C}$ of complex numbers, unless otherwise stated. The *vector addition*, denoted by $u+v$, is an operation between elements u and v of V, while the *scalar multiplication*, written as λv, is an operation between elements λ of $\mathbb{F}$ and v of V. We say that V is a *vector space* over $\mathbb{F}$ if the following hold:

1. $u + v \in V$ for all $u, v \in V$.
2. $\lambda v \in V$ for all $\lambda \in \mathbb{F}$ and $v \in V$.
3. $u + v = v + u$ for all $u, v \in V$.
4. $(u + v) + w = u + (v + w)$ for all $u, v, w \in V$.
5. There is an element $0 \in V$ such that $v + 0 = v$ for all $v \in V$.
6. For each $v \in V$ there exists an element $-v \in V$ such that $v+(-v) = 0$.
7. $\lambda(u + v) = \lambda u + \lambda v$ for all $\lambda \in \mathbb{F}$ and $u, v \in V$.
8. $(\lambda + \mu)v = \lambda v + \mu v$ for all $\lambda, \mu \in \mathbb{F}$ and $v \in V$.
9. $(\lambda\mu)v = \lambda(\mu v)$ for all $\lambda, \mu \in \mathbb{F}$ and $v \in V$.
10. $1v = v$ for all $v \in V$.

Some Important Vector Spaces.

- The xy-plane (also called the *Cartesian plane*) is a vector space over $\mathbb{R}$. Here we view the xy-plane as the set of arrows (directed line segments) in the plane, all with initial point O, the origin. Define the addition by the parallelogram law, which states that for two vectors u and v, the sum $u + v$ is the vector defined by the diagonal of the parallelogram with u and v as adjacent sides. Define λv to be the vector whose length is $|\lambda|$ times the length of v, pointing in the same direction as v if $\lambda \geq 0$ and otherwise pointing in the opposite direction. Note that the extreme case where the terminal point of the arrow coincides with O gives the zero vector for which the length of the arrow is 0 and any direction may be regarded as its direction. This vector space can be identified with the space $\mathbb{R}^2$ defined below.

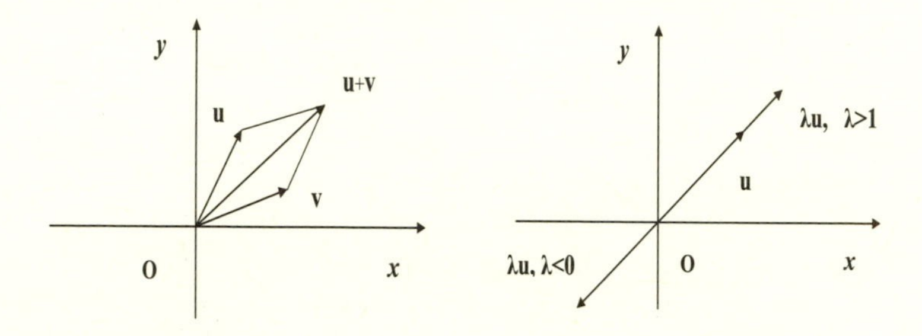

Figure 1.1: Vector addition and scalar multiplication

- The three-dimensional vector space over $\mathbb{R}$ consisting of all arrows starting from the origin in the ordinary three-dimensional space, with vector addition and scalar multiplication similarly defined (by the parallelogram rule) as above for the xy-plane. This space can be identified with the space $\mathbb{R}^3$ defined below.

  The spaces $\mathbb{R}^2$ and $\mathbb{R}^3$ will help the reader understand and visualize many concepts of vector spaces.

- $\mathbb{F}^n$ is a vector space over a field $\mathbb{F}$, where n is a positive integer and

$$\mathbb{F}^n = \left\{ \begin{pmatrix} x_1 \\ x_2 \\ \vdots \\ x_n \end{pmatrix} \;\middle|\; x_1, x_2, \ldots, x_n \in \mathbb{F} \right\}.$$

Here addition and scalar multiplication are defined, respectively, by

$$\begin{pmatrix} x_1 \\ x_2 \\ \vdots \\ x_n \end{pmatrix} + \begin{pmatrix} y_1 \\ y_2 \\ \vdots \\ y_n \end{pmatrix} = \begin{pmatrix} x_1 + y_1 \\ x_2 + y_2 \\ \vdots \\ x_n + y_n \end{pmatrix}, \quad \lambda \begin{pmatrix} x_1 \\ x_2 \\ \vdots \\ x_n \end{pmatrix} = \begin{pmatrix} \lambda x_1 \\ \lambda x_2 \\ \vdots \\ \lambda x_n \end{pmatrix}.$$

In particular, $\mathbb{R}^n$ and $\mathbb{C}^n$ are vector spaces over $\mathbb{R}$ and $\mathbb{C}$, respectively.

Note: In the context of vector spaces, it usually makes no difference whether to write a vector in $\mathbb{F}^n$ as a row or a column. So sometimes we may write the vectors in $\mathbb{F}^n$ as rows $(x_1, x_2, \ldots, x_n)$ for convenience. However, when a matrix-vector product Ax is involved, it is clear from the context that x has to be a column vector.

- $M_{m\times n}(\mathbb{F})$ over $\mathbb{F}$, where m and n are positive integers and $M_{m\times n}(\mathbb{F})$ is the collection of all $m \times n$ matrices over a scalar field $\mathbb{F}$. An $m \times n$ *matrix* over $\mathbb{F}$ is an array of m rows and n columns:

$$A = \begin{pmatrix} a_{11} & a_{12} & \cdots & a_{1n} \\ a_{21} & a_{22} & \cdots & a_{2n} \\ \cdots & \cdots & \cdots & \cdots \\ a_{m1} & a_{m2} & \cdots & a_{mn} \end{pmatrix}.$$

The notation $A = (a_{ij})_{m\times n}$ or $A = (a_{ij})$ is sometimes used for simplicity. If $m = n$, we often write $M_n(\mathbb{F})$ for $M_{m\times n}(\mathbb{F})$. Two matrices are *equal* if they have the same size and same corresponding entries.

The *addition* of two $m \times n$ matrices is defined by adding the corresponding entries, and the *scalar multiplication* of a matrix by a scalar is obtained by multiplying every entry of the matrix by the scalar. In symbols, if $A = (a_{ij})$, $B = (b_{ij}) \in M_{m\times n}(\mathbb{F})$, and $\lambda \in \mathbb{F}$, then

$$A + B = (a_{ij} + b_{ij}), \qquad \lambda A = (\lambda a_{ij}).$$

Note: If $m = 1$ or $n = 1$, $M_{m\times n}(\mathbb{F})$ can be identified with $\mathbb{F}^n$ or $\mathbb{F}^m$.

Matrices can also be multiplied when they have appropriate sizes. Let A be a $p \times n$ matrix and B be an $n \times q$ matrix. The *matrix product* AB of A and B is a $p \times q$ matrix whose (i, j)-entry is given by $a_{i1}b_{1j} + a_{i2}b_{2j} + \cdots + a_{in}b_{nj}$, $i = 1, 2, \ldots, p$, $j = 1, 2, \ldots, q$. So, to add two matrices, the matrices must have the same size, while to multiply two matrices, the number of columns of the first matrix must equal the number of rows of the second matrix. Note that even though AB is well defined, BA may not be; moreover $AB \neq BA$ in general.

The *zero matrix* of size $m \times n$, abbreviated to 0 when the size is clear or not important, is the $m \times n$ matrix all whose entries are 0. The *identity matrix* of size $n \times n$, shortened to I_n or simply I, is the n-square matrix whose main diagonal entries are all 1 and off-diagonal entries are all 0. A square matrix A is said to be *invertible,* or *nonsingular*, if there exists a matrix B such that $AB = BA = I$. Such a matrix is called the *inverse* of A and denoted by A^{-1}.

Besides the properties on addition and scalar multiplication, as a vector space, $M_{m\times n}(\mathbb{F})$ satisfies the following:

(a) $0A = A0 = 0$.

(b) $AI = IA = A$.

(c) $(AB)C = A(BC)$.

(d) $A(B + C) = AB + AC$.

(e) $(A + B)C = AC + BC$.

(f) $k(AB) = (kA)B = A(kB)$, where k is a scalar.

For an $m \times n$ matrix $A = (a_{ij})$, we can associate an $n \times m$ matrix to A by converting the rows of A to columns; that is, equivalently, the (i, j)-entry of the resulting matrix is a_{ji}. Such a matrix is called the *transpose* of A and denoted by A^t. If A is a complex matrix, as it usually is in this book, we define the *conjugate* of A by taking the conjugate of each entry: $\bar{A} = (\bar{a}_{ij})$. We write A^* for the *conjugate transpose* of A, namely, $A^* = (\bar{A})^t$. The following properties hold:

(i) $(A^t)^t = A$; $(A^*)^* = A$.

(ii) $(A + B)^t = A^t + B^t$; $(A + B)^* = A^* + B^*$.

(iii) $(AB)^t = B^t A^t$; $(AB)^* = B^* A^*$.

(iv) $(kA)^t = kA^t$; $(kA)^* = \bar{k}A^*$, where k is a scalar.

Let A be a matrix. A *submatrix* of A is a matrix that consists of the entries of A lying in certain rows and columns of A. For example, let

$$A = \begin{pmatrix} 1 & 2 & 3 \\ 4 & 5 & 6 \\ 7 & 8 & 9 \end{pmatrix}, \quad B = \begin{pmatrix} 2 & 3 \\ 5 & 6 \end{pmatrix}, \quad C = \begin{pmatrix} 5 & 6 \\ 8 & 9 \end{pmatrix}.$$

B is a submatrix of A lying in rows 1 and 2 and columns 2 and 3 of A, and C is a submatrix of A obtained by deleting the first row and the first column of A. Sometimes it is useful and convenient to *partition* a matrix into submatrices. For instance, we may write

$$A = \begin{pmatrix} 1 & 2 & 3 \\ 4 & 5 & 6 \\ 7 & 8 & 9 \end{pmatrix} = \begin{pmatrix} X & B \\ U & V \end{pmatrix},$$

where

$$X = \begin{pmatrix} 1 \\ 4 \end{pmatrix}, \quad B = \begin{pmatrix} 2 & 3 \\ 5 & 6 \end{pmatrix}, \quad U = (7), \quad V = (8,\ 9).$$

Let $A = (a_{ij})$ be an $n \times n$ complex matrix. The matrix A is said to be *Hermitian* if $A^* = A$; *symmetric* if $A^t = A$; *skew-Hermitian* if $A^* = -A$; *normal* if $A^*A = AA^*$; *upper-triangular* if $a_{ij} = 0$ whenever $i > j$; *lower-triangular* if $a_{ij} = 0$ whenever $i < j$; *diagonal* if $a_{ij} = 0$ whenever $i \neq j$, written as $A = \operatorname{diag}(a_{11}, a_{22}, \dots, a_{nn})$; *unitary* if $A^*A = AA^* = I$; and *real orthogonal* if A is real and $A^tA = AA^t = I$.

- $\mathbb{P}_n[x]$ over a field $\mathbb{F}$, where n is a positive integer and $\mathbb{P}_n[x]$ is the set of all polynomials of degree less than n with coefficients from $\mathbb{F}$. A constant polynomial $p(x) = a_0$ is said to have degree 0 if $a_0 \neq 0$, or degree $-\infty$ if $a_0 = 0$. The *addition* and *scalar multiplication* are defined for $p,\ q \in \mathbb{P}_n[x]$, and $\lambda \in \mathbb{F}$ by

$$\begin{aligned} (p+q)(x) &= p(x) + q(x) \\ &= (a_{n-1} + b_{n-1})x^{n-1} + \cdots + (a_1 + b_1)x + (a_0 + b_0), \end{aligned}$$

where

$$p(x) = a_{n-1}x^{n-1} + \cdots + a_1x + a_0,\ q(x) = b_{n-1}x^{n-1} + \cdots + b_1x + b_0$$

and

$$(\lambda p)(x) = \lambda\,(p(x)) = (\lambda a_{n-1})x^{n-1} + \cdots + (\lambda a_1)x + (\lambda a_0).$$

Denote by $\mathbb{P}[x]$ the collection of all polynomials of any finite degree with coefficients from $\mathbb{F}$. Then $\mathbb{P}[x]$ is a vector space over $\mathbb{F}$ with respect to the above operations for polynomials.

- $\mathcal{C}[a,b]$ over $\mathbb{R}$, where $\mathcal{C}[a,b]$ is the set of all real-valued continuous functions on the interval $[a,b]$. Functions are added and multiplied in the usual way, i.e., if f and g are continuous functions on $[a,b]$, then $(f+g)(x) = f(x) + g(x)$ and $(\lambda f)(x) = \lambda f(x)$, where $\lambda \in \mathbb{R}$. $\mathcal{C}(\mathbb{R})$ denotes the vector space of real-valued continuous functions on $\mathbb{R}$.

Linear Dependence. Let $v_1, v_2, \ldots, v_n$ be vectors of a vector space V over a field $\mathbb{F}$ and let $\lambda_1, \lambda_2, \ldots, \lambda_n$ be scalars from $\mathbb{F}$. Then the vector

$$v = \lambda_1 v_1 + \lambda_2 v_2 + \cdots + \lambda_n v_n$$

is called a *linear combination* of the vectors $v_1, v_2, \ldots, v_n$, and the scalars $\lambda_1, \lambda_2, \ldots, \lambda_n$ are called the *coefficients* of the linear combination. If all the coefficients are zero, then $v = 0$. There may exist a linear combination of the vectors $v_1, v_2, \ldots, v_n$ that equals zero even though the coefficients $\lambda_1, \lambda_2, \ldots, \lambda_n$ are not all zero. In this case, we say that the vectors $v_1, v_2, \ldots, v_n$ are *linearly dependent*. In other words, the vectors $v_1, v_2, \ldots, v_n$ are linearly dependent if and only if there exist scalars λ_1, λ_2, $\ldots$, λ_n, not all zero, such that

$$\lambda_1 v_1 + \lambda_2 v_2 + \cdots + \lambda_n v_n = 0. \tag{1.1}$$

The vectors $v_1, v_2, \ldots, v_n$ are *linearly independent* if they are not linearly dependent, i.e., $v_1, v_2, \ldots, v_n$ are linearly independent if (1.1) holds only when all the coefficients $\lambda_1, \lambda_2, \ldots, \lambda_n$ are zero. The zero vector 0 itself is linearly dependent because $\lambda 0 = 0$ for any nonzero scalar λ.

Dimension and Bases. The largest number of linearly independent vectors in a vector space V is called the *dimension* of V, written as $\dim V$. If that is a finite number n, we define $\dim V = n$ and say V is finite dimensional. If there are arbitrarily large independent sets in V, we say $\dim V$ is infinite and V is infinite dimensional. For the finite dimensional case, if there exist n vectors in V that are linearly independent and any $n+1$ vectors in V are linearly dependent, then $\dim V = n$. In this case, any set of n linearly independent vectors is called a *basis* for the vector space V. The vector space of one element, zero, is said to have dimension 0 with no basis. Note that the dimension of a vector space also depends on the underlying number field, $\mathbb{F}$, of the vector space. Unless otherwise stated, we assume throughout the book that *vector spaces are finite dimensional.*

For the scalar field $\mathbb{F}$, the dimension of the vector space $\mathbb{F}^n$ is n, and the vectors $e_1 = (1, 0, 0, \ldots, 0), e_2 = (0, 1, 0, \ldots, 0), \ldots, e_n = (0, 0, \ldots, 0, 1)$ (sometimes written as column vectors) are a basis for $\mathbb{F}^n$, refereed to as the *standard basis* for $\mathbb{F}^n$.

Let $\{\alpha_1, \alpha_2, \ldots, \alpha_n\}$ be a basis of the vector space V, and let v be any vector in V. Since $v, \alpha_1, \alpha_2, \ldots, \alpha_n$ are linearly dependent ($n+1$ vectors), there are scalars $\lambda, \lambda_1, \lambda_2, \ldots, \lambda_n$, not all zero, such that

$$\lambda v + \lambda_1 \alpha_1 + \lambda_2 \alpha_2 + \cdots + \lambda_n \alpha_n = 0.$$

Since $\alpha_1, \alpha_2, \ldots, \alpha_n$ are linearly dependent, we see $\lambda \neq 0$. Thus

$$v = x_1 \alpha_1 + x_2 \alpha_2 + \cdots + x_n \alpha_n,$$

where $x_i = -\lambda_i/\lambda$, $i = 1, 2, \ldots, n$. Again due to the linear independence of $\alpha_1, \alpha_2, \ldots, \alpha_n$, such an expression of v as a linear combination of $\alpha_1, \alpha_2, \ldots, \alpha_n$ must be unique. We call the n-tuple $(x_1, x_2, \ldots, x_n)$ the *coordinate* of v under the (ordered) basis $\alpha_1, \alpha_2, \ldots, \alpha_n$; sometimes we also say that $x_1, x_2, \ldots, x_n$ are the *coordinates* of v under the basis $\{\alpha_1, \alpha_2, \ldots, \alpha_n\}$.

Subspace. Let V be a vector space over a field $\mathbb{F}$ and W be a nonempty subset of V. If W is also a vector space over $\mathbb{F}$ under the same vector addition and scalar multiplication of V, then W is said to be a *subspace* of V. One may check that W is a subspace of V if and only if W is closed under the operations of V; that is, (i) if u, $v \in W$ then $u + v \in W$ and (ii) if $v \in W$ and $\lambda \in \mathbb{F}$ then $\lambda v \in W$. It follows that, to be a subspace, W must contain the zero vector 0 of V. $\{0\}$ and V are trivial subspaces of V. A subspace W of V is called a *proper subspace* if $W \neq V$.

Let W_1 and W_2 be subspaces of a vector space V. The *intersection*

$$W_1 \cap W_2 = \{ v \mid v \in W_1 \text{ and } v \in W_2 \}$$

is also a subspace of V and so is the *sum*

$$W_1 + W_2 = \{ w_1 + w_2 \mid w_1 \in W_1 \text{ and } w_2 \in W_2 \}.$$

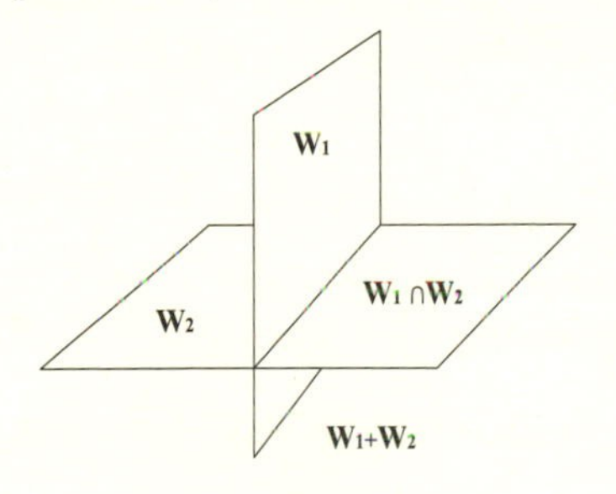

Figure 1.2: Sum of subspaces

The sum $W_1 + W_2$ is called a *direct sum*, denoted by $W_1 \oplus W_2$, if every element v in $W_1 + W_2$ can be uniquely written as $v = w_1 + w_2$, where $w_1 \in W_1$, $w_2 \in W_2$; that is, if $v = v_1 + v_2$, where $v_1 \in W_1$, $v_2 \in W_2$, then $w_1 = v_1$ and $w_2 = v_2$. In particular, if $0 = w_1 + w_2$, then $w_1 = w_2 = 0$.

Let S be a nonempty subset of V. The subspace Span(S) is defined to consist of all possible (finite) linear combinations of the elements of S. In particular, if S is a finite set, say, $S = \{v_1, v_2, \ldots, v_k\}$, then

$$\operatorname{Span}(S) = \{\, \lambda_1 v_1 + \lambda_2 v_2 + \cdots + \lambda_k v_k \mid \lambda_1, \lambda_2, \ldots, \lambda_k \in \mathbb{F} \,\}.$$

For any nonempty S, Span(S) is a subspace of the vector space V. We say the subspace Span(S) is *spanned* by S, or *generated* by S.

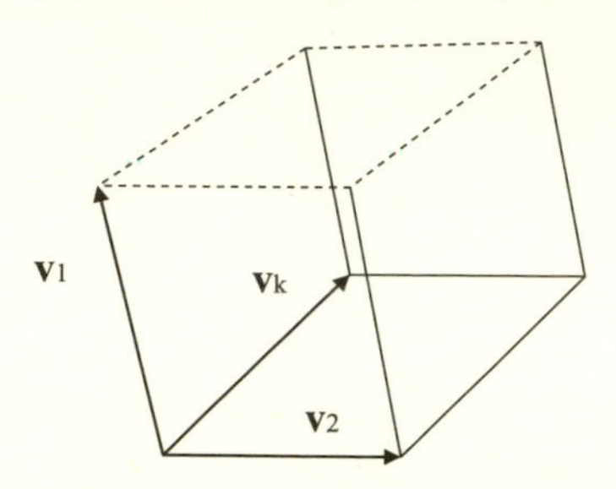

Figure 1.3: Subspace spanned by vectors

Given an $m \times n$ matrix A over a scalar field $\mathbb{F}$, there are three important spaces associated to A. The space spanned by the rows of A is a subspace of $\mathbb{F}^n$, called *row space* of A. The space spanned by the columns of A is a subspace of $\mathbb{F}^m$, called the *column space* of A. The column space of a matrix A is also known as the *image* or *range* of A, denoted by $\operatorname{Im} A$; this origins from A being viewed as the mapping from $\mathbb{F}^n$ to $\mathbb{F}^m$ defined by $x \mapsto Ax$. Both terms and notations are in practical use. Thus

$$\operatorname{Im} A = \{\, Ax \mid x \in \mathbb{F}^n \,\}.$$

All solutions to the equation system $Ax = 0$ form a subspace of $\mathbb{F}^n$. This space is called *null space* or *kernel* of A and symbolized by $\operatorname{Ker} A$. So

$$\operatorname{Ker} A = \{\, x \in \mathbb{F}^n \mid Ax = 0 \,\}.$$

Dimension Identity. Let W_1, W_2 be subspaces of a vector space V. Then

$$\dim W_1 + \dim W_2 = \dim(W_1 + W_2) + \dim(W_1 \cap W_2).$$

Chapter 1 Problems

1.1 Let $\mathbb{C}$, $\mathbb{R}$, and $\mathbb{Q}$ be the fields of complex, real, and rational numbers, respectively. Determine whether each of the following is a vector space. Find the dimension and a basis for each that is a vector space.

(a) $\mathbb{C}$ over $\mathbb{C}$.

(b) $\mathbb{C}$ over $\mathbb{R}$.

(c) $\mathbb{R}$ over $\mathbb{C}$.

(d) $\mathbb{R}$ over $\mathbb{Q}$.

(e) $\mathbb{Q}$ over $\mathbb{R}$.

(f) $\mathbb{Q}$ over $\mathbb{Z}$, where $\mathbb{Z}$ is the set of all integers.

(g) $\mathbb{S} = \{ a + b\sqrt{2} + c\sqrt{5} \mid a, b, c \in \mathbb{Q} \}$ over $\mathbb{Q}$, $\mathbb{R}$, or $\mathbb{C}$.

1.2 Consider $\mathbb{R}^2$ over $\mathbb{R}$. Give an example of a subset of $\mathbb{R}^2$ that is

(a) closed under addition but not under scalar multiplication;

(b) closed under scalar multiplication but not under addition.

1.3 Let $V = \{ (x, y) \mid x, y \in \mathbb{C} \}$. Under the standard addition and scalar multiplication for ordered pairs of complex numbers, is V a vector space over $\mathbb{C}$? Over $\mathbb{R}$? Over $\mathbb{Q}$? If so, find the dimension of V.

1.4 Why does a vector space V over $\mathbb{F}$ ($= \mathbb{C}$, $\mathbb{R}$, or $\mathbb{Q}$) have either one element or infinitely many elements? Given $v \in V$, is it possible to have two distinct vectors u, w in V such that $u+v=0$ and $w+v=0$?

1.5 Let V be the collection of all real ordered pairs in which the second number is twice the first one; that is, $V = \{ (x, y) \mid y = 2x, \ x \in \mathbb{R} \}$. If the addition and multiplication are defined, respectively, to be

$$(x_1, y_1) + (x_2, y_2) = (x_1 + x_2, y_1 + y_2), \quad \lambda \cdot (x, y) = (\lambda x, \lambda y),$$

show that V is a vector space over $\mathbb{R}$ with respect to the operations. Is V also a vector space with respect to the above addition and the scalar multiplication defined instead by $\lambda \odot (x, y) = (\lambda x, 0)$? [Note: The reason for the use of the symbol $\odot$ instead of $\cdot$ is to avoid confusion when the two operations are discussed in the same problem.]

1.6 Let $\mathbb{H}$ be the collection of all 2×2 complex matrices of the form

$$\begin{pmatrix} a & b \\ -\bar{b} & \bar{a} \end{pmatrix}.$$

Show that $\mathbb{H}$ is a vector space (under the usual matrix addition and scalar multiplication) over $\mathbb{R}$. Is $\mathbb{H}$ also a vector space over $\mathbb{C}$?

1.7 Let $\mathbb{R}^+$ be the set of all positive real numbers. Show that $\mathbb{R}^+$ is a vector space over $\mathbb{R}$ under the addition

$$x \boxplus y = xy, \quad x,\ y \in \mathbb{R}^+$$

and the scalar multiplication

$$a \boxdot x = x^a, \quad x \in \mathbb{R}^+,\ a \in \mathbb{R}.$$

Find the dimension of the vector space. Is $\mathbb{R}^+$ also a vector space over $\mathbb{R}$ if the scalar multiplication is instead defined as

$$a \boxtimes x = a^x, \quad x \in \mathbb{R}^+,\ a \in \mathbb{R}?$$

1.8 Let $\{\alpha_1, \alpha_2, \ldots, \alpha_n\}$ be a basis of an n-dimensional vector space V. Show that $\{\lambda_1\alpha_1, \lambda_2\alpha_2, \ldots, \lambda_n\alpha_n\}$ is also a basis of V for any nonzero scalars $\lambda_1, \lambda_2, \ldots, \lambda_n$. If the coordinate of a vector v under the basis $\{\alpha_1, \alpha_2, \ldots, \alpha_n\}$ is $x = (x_1, x_2, \ldots, x_n)$, what is the coordinate of v under $\{\lambda_1\alpha_1, \lambda_2\alpha_2, \ldots, \lambda_n\alpha_n\}$? What are the coordinates of $w = \alpha_1 + \alpha_2 + \cdots + \alpha_n$ under $\{\alpha_1, \alpha_2, \ldots, \alpha_n\}$ and $\{\lambda_1\alpha_1, \lambda_2\alpha_2, \ldots, \lambda_n\alpha_n\}$?

1.9 Let $v_1, v_2, \ldots, v_k$ be vectors in a vector space V. State what is meant for $\{v_1, v_2, \ldots, v_k\}$ to be a basis of V using (i) the words "span" and "independent"; (ii) instead the phrase "linear combination."

1.10 Consider k vectors in $\mathbb{R}^n$ and answer the three questions in cases of $k < n$, $k = n$, and $k > n$: (i) Are the vectors linearly independent? (ii) Do they span $\mathbb{R}^n$? (iii) Do they form a basis for $\mathbb{R}^n$?

1.11 Let $\{\alpha_1,\ \alpha_2,\ \alpha_3\}$ be a basis for $\mathbb{R}^3$ and let $\alpha_4 = -\alpha_1 - \alpha_2 - \alpha_3$. Show that every vector v in $\mathbb{R}^3$ can be written as $v = a_1\alpha_1 + a_2\alpha_2 + a_3\alpha_3 + a_4\alpha_4$, where a_1, a_2, a_3, a_4 are unique real numbers such that $a_1 + a_2 + a_3 + a_4 = 0$. Generalize this to a vector space of dimension n.

1.12 Show that $\{1, (x-1), (x-1)(x-2)\}$ is a basis of $\mathbb{P}_3[x]$ and that $W = \{p(x) \in \mathbb{P}_3[x] \mid p(1) = 0\}$ is a subspace of $\mathbb{P}_3[x]$. Find $\dim W$.

1.13 Answer true or false:

(a) $\{(x, y) \mid x^2 + y^2 = 0,\ x,\ y \in \mathbb{R}\}$ is a subspace of $\mathbb{R}^2$.

(b) $\{(x, y) \mid x^2 + y^2 \leq 1,\ x,\ y \in \mathbb{R}\}$ is a subspace of $\mathbb{R}^2$.

(c) $\{(x, y) \mid x^2 + y^2 = 0,\ x,\ y \in \mathbb{C}\}$ is a subspace of $\mathbb{C}^2$.

(d) $\{(x, y) \mid x^2 - y^2 = 0,\ x,\ y \in \mathbb{R}\}$ is a subspace of $\mathbb{R}^2$.

(e) $\{(x, y) \mid x - y = 0,\ x,\ y \in \mathbb{R}\}$ is a subspace of $\mathbb{R}^2$.

(f) $\{(x, y) \mid x + y = 0,\ x,\ y \in \mathbb{R}\}$ is a subspace of $\mathbb{R}^2$.

(g) $\{(x, y) \mid xy = 0,\ x,\ y \in \mathbb{R}\}$ is a subspace of $\mathbb{R}^2$.

(h) $\{(x, y) \mid xy \geq 0,\ x,\ y \in \mathbb{R}\}$ is a subspace of $\mathbb{R}^2$.

(i) $\{(x, y) \mid x > 0,\ y > 0\}$ is a subspace of $\mathbb{R}^2$.

(j) $\{(x, y) \mid x,\ y \text{ are integers}\}$ is a subspace of $\mathbb{R}^2$.

(k) $\{(x, y) \mid x/y = 1,\ x,\ y \in \mathbb{R}\}$ is a subspace of $\mathbb{R}^2$.

(l) $\{(x, y) \mid y = 3x,\ x,\ y \in \mathbb{R}\}$ is a subspace of $\mathbb{R}^2$.

(m) $\{(x, y) \mid x - y = 1,\ x,\ y \in \mathbb{R}\}$ is a subspace of $\mathbb{R}^2$.

1.14 Consider $\mathbb{P}_n[x]$ and $\mathbb{P}[x]$ over $\mathbb{R}$. Answer true or false:

(a) $\{p(x) \mid p(x) = ax + b,\ a,\ b \in \mathbb{R}\}$ is a subspace of $\mathbb{P}_3[x]$.

(b) $\{p(x) \mid p(x) = ax^2,\ a \in \mathbb{R}\}$ is a subspace of $\mathbb{P}_3[x]$.

(c) $\{p(x) \mid p(x) = a + x^2,\ a \in \mathbb{R}\}$ is a subspace of $\mathbb{P}_3[x]$.

(d) $\{p(x) \mid p(x) \in \mathbb{P}[x] \text{ has degree } 3\}$ is a subspace of $\mathbb{P}[x]$.

(e) $\{p(x) \mid p(0) = 0,\ p(x) \in \mathbb{P}[x]\}$ is a subspace of $\mathbb{P}[x]$.

(f) $\{p(x) \mid p(0) = 1,\ p(x) \in \mathbb{P}[x]\}$ is a subspace of $\mathbb{P}[x]$.

(g) $\{p(x) \mid 2p(0) = p(1),\ p(x) \in \mathbb{P}[x]\}$ is a subspace of $\mathbb{P}[x]$.

(h) $\{p(x) \mid p(x) \geq 0,\ p(x) \in \mathbb{P}[x]\}$ is a subspace of $\mathbb{P}[x]$.

(i) $\{p(x) \mid p(-x) = p(x),\ p(x) \in \mathbb{P}[x]\}$ is a subspace of $\mathbb{P}[x]$.

(j) $\{p(x) \mid p(-x) = -p(x),\ p(x) \in \mathbb{P}[x]\}$ is a subspace of $\mathbb{P}[x]$.

1.15 Consider the real vector space $\mathbb{R}^4$. Let

$$\alpha_1 = (1, -3, 0, 2), \quad \alpha_2 = (-2, 1, 1, 1), \quad \alpha_3 = (-1, -2, 1, 3).$$

Determine whether α_1, α_2, and α_3 are linearly dependent. Find the dimension and a basis for the subspace $\operatorname{Span}\{\alpha_1, a_2, a_3\}$.

1.16 Let V be the subspace of $\mathbb{R}^4$ spanned by the 4-tuples

$$\alpha_1 = (1, 2, 3, 4),\ \alpha_2 = (2, 3, 4, 5),\ \alpha_3 = (3, 4, 5, 6),\ \alpha_4 = (4, 5, 6, 7).$$

Find a basis of V and $\dim V$.

1.17 Let α_1, α_2, α_3, and α_4 be linearly independent. Answer true or false:

(a) $\alpha_1 + \alpha_2$, $\alpha_2 + \alpha_3$, $\alpha_3 + \alpha_4$, $\alpha_4 + \alpha_1$ are linearly independent.

(b) $\alpha_1 - \alpha_2$, $\alpha_2 - \alpha_3$, $\alpha_3 - \alpha_4$, $\alpha_4 - \alpha_1$ are linearly independent.

(c) $\alpha_1 + \alpha_2$, $\alpha_2 + \alpha_3$, $\alpha_3 + \alpha_4$, $\alpha_4 - \alpha_1$ are linearly independent.

(d) $\alpha_1 + \alpha_2$, $\alpha_2 + \alpha_3$, $\alpha_3 - \alpha_4$, $\alpha_4 - \alpha_1$ are linearly independent.

1.18 Let α_1, α_2, α_3 be linearly independent. For what value of k are the vectors $\alpha_2 - \alpha_1$, $k\alpha_3 - \alpha_2$, $\alpha_1 - \alpha_3$ linearly independent?

1.19 If α_1, α_2, α_3 are linearly dependent and α_2, α_3, α_4 are linearly independent, show that (i) α_1 is a linear combination of α_2 and α_3, and (ii) α_4 is not a linear combination of α_1, α_2, and α_3.

1.20 Show that $\alpha_1 = (1, 1, 0)$, $\alpha_2 = (1, 0, 1)$, and $\alpha_3 = (0, 1, 1)$ form a basis for $\mathbb{R}^3$. Find the coordinates of the vectors $u = (2, 0, 0)$, $v = (1, 0, 0)$, and $w = (1, 1, 1)$ under the basis $\{\alpha_1, \alpha_2, \alpha_3\}$.

1.21 Let $W = \left\{ \left(\begin{smallmatrix} a & b \\ b & c \end{smallmatrix}\right) \mid a,\ b,\ c \in \mathbb{R} \right\}$. Show that W is a subspace of $M_2(\mathbb{R})$ over $\mathbb{R}$ and that the following matrices form a basis for W:

$$\begin{pmatrix} 1 & 0 \\ 0 & 0 \end{pmatrix}, \quad \begin{pmatrix} 0 & 1 \\ 1 & 0 \end{pmatrix}, \quad \begin{pmatrix} 0 & 0 \\ 0 & 1 \end{pmatrix}.$$

Find the coordinates of the matrix $\left(\begin{smallmatrix} 1 & -2 \\ -2 & 3 \end{smallmatrix}\right)$ under the basis.

1.22 Consider $\mathbb{P}_n[x]$ and $\mathbb{P}[x]$ over $\mathbb{R}$.

(a) Show that $\mathbb{P}_n[x]$ is a vector space over $\mathbb{R}$ under the ordinary addition and scalar multiplication for polynomials.

(b) Show that $\{1, x, x^2, \ldots, x^{n-1}\}$ is a basis for $\mathbb{P}_n[x]$, and so is

$$\{1, (x-a), (x-a)^2, \ldots, (x-a)^{n-1}\}, \quad a \in \mathbb{R}.$$

(c) Find the coordinate of

$$f(x) = a_0 + a_1 x + \cdots + a_{n-1}x^{n-1} \in \mathbb{P}_n[x]$$

with respect to the basis

$$\{1, (x-a), (x-a)^2, \ldots, (x-a)^{n-1}\}.$$

(d) Let $a_1, a_2, \ldots, a_n \in \mathbb{R}$ be distinct. For $i = 1, 2, \ldots, n$, let

$$f_i(x) = (x-a_1)\cdots(x-a_{i-1})(x-a_{i+1})\cdots(x-a_n).$$

Show that $\{f_1(x), \ldots, f_n(x)\}$ is also a basis for $\mathbb{P}_n[x]$.

(e) Show that $W = \{f(x) \in \mathbb{P}_n[x] \mid f(1) = 0\}$ is a subspace of $\mathbb{P}_n[x]$. Find its dimension and a basis.

(f) Is $\mathbb{P}[x]$ a vector space over $\mathbb{R}$? Is it of finite dimension?

(g) Show that each $\mathbb{P}_n[x]$ is a proper subspace of $\mathbb{P}[x]$.

1.23 Let $C(\mathbb{R})$ be the vector space of all real-valued continuous functions over $\mathbb{R}$ with addition $(f+g)(x) = f(x) + g(x)$ and scalar multiplication $(rf)(x) = rf(x)$, $r \in \mathbb{R}$. Show that $\sin x$ and $\cos x$ are linearly independent and that the vector space generated by $\sin x$ and $\cos x$

$$\text{Span}\{\sin x, \cos x\} = \{a \sin x + b \cos x \mid a, b \in \mathbb{R}\}$$

is contained in the solution set to the differential equation

$$y'' + y = 0.$$

Are $\sin^2 x$ and $\cos^2 x$ linearly independent? How about 1, $\sin^2 x$, and $\cos^2 x$? Find $\mathbb{R} \cap \text{Span}\{\sin x, \cos x\}$ and $\mathbb{R} \cap \text{Span}\{\sin^2 x, \cos^2 x\}$.

1.24 Let $t \in \mathbb{R}$. Discuss the linear independence of the vectors over $\mathbb{R}$:

$$\alpha_1 = (1,\ 1,\ 0), \quad \alpha_2 = (1,\ 3,\ -1), \quad \alpha_3 = (5,\ 3,\ t).$$

1.25 Let V be a finite dimensional vector space and S be a subspace of V. Show that

(a) $\dim S \leq \dim V$.

(b) $\dim S = \dim V$ if and only if $S = V$.

(c) Every basis for S is contained in some basis for V.

(d) A basis for V need not contain a basis for S.

1.26 Consider the vector space $\mathbb{R}^n$ over $\mathbb{R}$ with the usual operations.

(a) Show that

$$e_1 = \begin{pmatrix} 1 \\ 0 \\ 0 \\ \vdots \\ 0 \end{pmatrix}, \quad e_2 = \begin{pmatrix} 0 \\ 1 \\ 0 \\ \vdots \\ 0 \end{pmatrix}, \quad \dots, \quad e_n = \begin{pmatrix} 0 \\ 0 \\ 0 \\ \vdots \\ 1 \end{pmatrix}$$

and

$$\epsilon_1 = \begin{pmatrix} 1 \\ 0 \\ 0 \\ \vdots \\ 0 \end{pmatrix}, \quad \epsilon_2 = \begin{pmatrix} 1 \\ 1 \\ 0 \\ \vdots \\ 0 \end{pmatrix}, \quad \dots, \quad \epsilon_n = \begin{pmatrix} 1 \\ 1 \\ 1 \\ \vdots \\ 1 \end{pmatrix}$$

form two bases. Are they also bases for $\mathbb{C}^n$ over $\mathbb{C}$? over $\mathbb{R}$?

(b) Find a matrix A such that $A(\epsilon_1, \epsilon_2, \dots, \epsilon_n) = (e_1, e_2, \dots, e_n)$.

(c) Find a matrix B such that $(\epsilon_1, \epsilon_2, \dots, \epsilon_n) = B(e_1, e_2, \dots, e_n)$.

(d) If $v \in \mathbb{R}^n$ has the coordinate $(1, 2, \dots, n)$ on the basis $\{e_1, e_2, \dots, e_n\}$, what is the coordinate of v under $\{\epsilon_1, \epsilon_2, \dots, \epsilon_n\}$?

(e) Why are any $n+1$ vectors in $\mathbb{R}^n$ linearly dependent over $\mathbb{R}$?

(f) Find $n+1$ vectors in $\mathbb{C}^n$ that are linearly independent over $\mathbb{R}$.

1.27 Let $\{\alpha_1, \alpha_2, \dots, \alpha_n\}$ be a basis of a vector space V, $n \geq 2$. Show that $\{\alpha_1, \alpha_1 + \alpha_2, \dots, \alpha_1 + \alpha_2 + \cdots + \alpha_n\}$ is also a basis of V. Is the set

$$\{\alpha_1 + \alpha_2,\ \alpha_2 + \alpha_3, \dots, \alpha_{n-1} + \alpha_n,\ \alpha_n + \alpha_1\}$$

a basis for V too? How about the converse?

1.28 Show that $\alpha = \{\alpha_1, \alpha_2, \alpha_3\}$ and $\beta = \{\beta_1, \beta_2, \beta_3\}$ are bases for $\mathbb{R}^3$:

$$\alpha_1 = \begin{pmatrix} 1 \\ 1 \\ 1 \end{pmatrix}, \quad \alpha_2 = \begin{pmatrix} 1 \\ 0 \\ -1 \end{pmatrix}, \quad \alpha_3 = \begin{pmatrix} 1 \\ 0 \\ 1 \end{pmatrix},$$

$$\beta_1 = \begin{pmatrix} 1 \\ 2 \\ 1 \end{pmatrix}, \quad \beta_2 = \begin{pmatrix} 2 \\ 3 \\ 4 \end{pmatrix}, \quad \beta_3 = \begin{pmatrix} 3 \\ 4 \\ 3 \end{pmatrix}.$$

Find the matrix from basis α to basis β; that is, a matrix A such that

$$(\beta_1, \beta_2, \beta_3) = (\alpha_1, \alpha_2, \alpha_3)A.$$

If a vector $u \in \mathbb{R}^3$ has coordinate $(2, 0, -1)$ under the basis α, what is the coordinate of u under β?

1.29 If $\alpha_1, \alpha_2, \ldots, \alpha_n$ are linearly independent in a vector space V and $\alpha_1, \alpha_2, \ldots, \alpha_n, \beta$ are linearly dependent, show that β can be uniquely expressed as a linear combination of $\alpha_1, \alpha_2, \ldots, \alpha_n$.

1.30 Show that the vectors $\alpha_1 (\neq 0), \alpha_2, \ldots, \alpha_n$ of a vector space V are linearly dependent if and only if there exists an integer k, $1 < k \leq n$, such that α_k is a linear combination of $\alpha_1, \alpha_2, \ldots, \alpha_{k-1}$.

1.31 Let V and W be vector spaces over $\mathbb{F}$. Denote by $V \times W$ the collection of all ordered pairs (v, w), where $v \in V$ and $w \in W$, and define

$$(v_1, w_1) + (v_2, w_2) = (v_1 + v_2, w_1 + w_2)$$

and

$$k(v, w) = (kv, kw), \quad k \in \mathbb{F}.$$

(a) Show that $V \times W$ is a vector space over $\mathbb{F}$.

(b) Show that if V and W are finite dimensional, so is $V \times W$.

(c) Find $\dim(V \times W)$, given that $\dim V = m$ and $\dim W = n$.

(d) Explain why $\mathbb{R} \times \mathbb{R}^2$ can be identified with $\mathbb{R}^3$.

(e) Find a basis for $\mathbb{R}^2 \times M_2(\mathbb{R})$.

(f) What is the dimension of $M_2(\mathbb{R}) \times M_2(\mathbb{R})$?

1.32 Answer true or false:

(a) If the zero vector 0 is one of the vectors $\alpha_1, \alpha_2, \ldots, \alpha_r$, then these vectors are linearly dependent.

(b) If $\alpha_1, \alpha_2, \ldots, \alpha_r$ are linearly independent and α_{r+1} is not a linear combination of $\alpha_1, \alpha_2, \ldots, \alpha_r$, then the vectors $\alpha_1, \alpha_2, \ldots, \alpha_r, \alpha_{r+1}$ are also linearly independent.

(c) If α is a linear combination of $\beta_1, \beta_2, \ldots, \beta_m$, and each β_i, $i = 1, 2, \ldots, m$, is a linear combination of $\gamma_1, \gamma_2, \ldots, \gamma_n$, then α is a linear combination of $\gamma_1, \gamma_2, \ldots, \gamma_n$.

(d) If $\alpha_1, \alpha_2, \ldots, \alpha_r$ are linearly independent, then no α_i is a linear combination of the other vectors. How about the converse?

(e) If $\alpha_1, \alpha_2, \ldots, \alpha_r$ are linearly dependent, then any one of these vectors is a linear combination of the other vectors.

(f) If β is not a linear combination of $\alpha_1, \alpha_2, \ldots, \alpha_r$, then $\beta, \alpha_1, \alpha_2, \ldots, \alpha_r$ are linearly independent.

(g) If any $r - 1$ vectors of $\alpha_1, \alpha_2, \ldots, \alpha_r$ are linearly independent, then $\alpha_1, \alpha_2, \ldots, \alpha_r$ are linearly independent.

(h) If $V = \operatorname{Span}\{\alpha_1, \alpha_2, \ldots, \alpha_n\}$ and if every α_i is a linear combination of no more than r vectors in $\{\alpha_1, \alpha_2, \ldots, \alpha_n\}$ excluding α_i, then $\dim V \leq r$.

1.33 Let U and V be subspaces of $\mathbb{R}^n$ spanned by vectors $\alpha_1, \alpha_2, \ldots, \alpha_p$ and $\beta_1, \beta_2, \ldots, \beta_q$, respectively. Let W be spanned by $\alpha_i + \beta_j$, $i = 1, 2, \ldots, p$, $j = 1, 2, \ldots, q$. If $\dim U = s$ and $\dim V = t$, show that

$$\dim W \leq \min\{\, n,\ s + t \,\}.$$

1.34 Let $\alpha_1, \alpha_2, \ldots, \alpha_r$ be linearly independent. If vector u is a linear combination of $\alpha_1, \alpha_2, \ldots, \alpha_r$, while vector v is not, show that the vectors $tu + v$, $\alpha_1, \ldots, \alpha_r$ are linearly independent for any scalar t.

1.35 Given a square matrix A, show that $V = \{\, X \mid AX = XA \,\}$, the set of the matrices commuting with A, is a vector space. Find all matrices that commute with A and find the dimension of the space, where

$$A = \begin{pmatrix} 1 & 0 & 0 \\ 0 & 1 & 0 \\ 3 & 1 & 2 \end{pmatrix}.$$

1.36 Find a basis and the dimension for each of the following vector spaces:

(a) $M_n(\mathbb{C})$ over $\mathbb{C}$.

(b) $M_n(\mathbb{C})$ over $\mathbb{R}$.

(c) $M_n(\mathbb{R})$ over $\mathbb{R}$.

(d) $H_n(\mathbb{C})$, $n \times n$ Hermitian matrices, over $\mathbb{R}$.

(e) $H_n(\mathbb{R})$, $n \times n$ real symmetric matrices, over $\mathbb{R}$.

(f) $S_n(\mathbb{C})$, $n \times n$ skew-Hermitian matrices, over $\mathbb{R}$.

(g) $S_n(\mathbb{R})$, $n \times n$ real skew-Hermitian matrices, over $\mathbb{R}$.

(h) $U_n(\mathbb{R})$, $n \times n$ real upper-triangular matrices, over $\mathbb{R}$.

(i) $L_n(\mathbb{R})$, $n \times n$ real lower-triangular matrices, over $\mathbb{R}$.

(j) $D_n(\mathbb{R})$, $n \times n$ real diagonal matrices, over $\mathbb{R}$.

(k) The space of all real polynomials in A over $\mathbb{R}$, where

$$A = \begin{pmatrix} 1 & 0 & 0 \\ 0 & \omega & 0 \\ 0 & 0 & \omega^2 \end{pmatrix}, \quad \omega = \frac{-1+\sqrt{3}\,i}{2}.$$

For example, $A^3 - A^2 + 5A + I$ is one of the polynomials.

Is $H_n(\mathbb{C})$ a subspace of $M_n(\mathbb{C})$ over $\mathbb{C}$? Is the set of $n \times n$ normal matrices a subspace of $M_n(\mathbb{C})$ over $\mathbb{C}$? Show that every $n \times n$ complex matrix is a sum of a Hermitian matrix and a skew-Hermitian matrix.

1.37 Find the space of matrices commuting with matrix A, where

(a) $A = I_n$.

(b) $A = \begin{pmatrix} 1 & 1 \\ 0 & 1 \end{pmatrix}$.

(c) $A = \begin{pmatrix} a & 0 \\ 0 & b \end{pmatrix}$, $a \neq b$.

(d) $A = \begin{pmatrix} 0 & 1 & 0 & 0 \\ 0 & 0 & 1 & 0 \\ 0 & 0 & 0 & 1 \\ 0 & 0 & 0 & 0 \end{pmatrix}$.

(e) A is an arbitrary $n \times n$ matrix.

1.38 Let $A \in M_{m\times n}(\mathbb{C})$ and $S(A) = \{ X \in M_{n\times p}(\mathbb{C}) \mid AX = 0 \}$. Show that $S(A)$ is a subspace of $M_{n\times p}(\mathbb{C})$ and that if $m = n$, then

$$S(A) \subseteq S(A^2) \subseteq \cdots \subseteq S(A^k) \subseteq S(A^{k+1}) \text{ for any positive integer } k.$$

Show further that this inclusion chain must terminate; that is,

$$S(A^r) = S(A^{r+1}) = S(A^{r+2}) = \cdots \text{ for some positive integer } r.$$

1.39 Denote by $\operatorname{Im} X$ the *column space* or *image* of matrix X. Let A be $m\times p$, B be $m\times q$. Show that the following statements are equivalent:

(a) $\operatorname{Im} A \subseteq \operatorname{Im} B$.

(b) The columns of A are linear combinations of the columns of B.

(c) $A = BC$ for some $q \times p$ matrix C.

1.40 Denote by $\operatorname{Ker} X$ the *null space* of matrix X. Let A be an $m \times n$ matrix over a field $\mathbb{F}$. Show each of the following statements.

(a) $\operatorname{Ker} A$ is a subspace of $\mathbb{F}^n$ and $\operatorname{Ker} A = \{0\}$ if and only if the columns of A are linearly independent. If the columns of A are linearly independent, are the rows of A necessarily linearly independent?

(b) If $m < n$, then $\operatorname{Ker} A \neq \{0\}$.

(c) $\operatorname{Ker} A \subseteq \operatorname{Ker} A^2$.

(d) $\operatorname{Ker}(A^*A) = \operatorname{Ker} A$.

(e) If $A = BC$, where B is $m \times m$ and C is $m \times n$, and if B is nonsingular, then $\operatorname{Ker} A = \operatorname{Ker} C$.

1.41 Let W_1 and W_2 be nontrivial subspaces of a vector space V; that is, neither $\{0\}$ nor V. Show that there exists an element $\alpha \in V$ such that $\alpha \notin W_1$ and $\alpha \notin W_2$. Show further that there exists a basis of V such that none of the vectors in the basis is contained in either W_1 or W_2. Is this true for more than two nontrivial subspaces?

1.42 Let $\{v_1, v_2, \ldots, v_n\}$ be a basis of a vector space V. Suppose W is a k-dimensional subspace of V, $1 < k < n$. Show that, for any subset $\{v_{i_1}, v_{i_2}, \ldots, v_{i_m}\}$ of $\{v_1, v_2, \ldots, v_n\}$, $m > n-k$, there exists a nonzero vector $w \in W$, which is a linear combination of $v_{i_1}, v_{i_2}, \ldots, v_{i_m}$.

1.43 Let W_1 and W_2 be subspaces of a vector space V and define the *sum*

$$W_1 + W_2 = \{ w_1 + w_2 \mid w_1 \in W_1,\ w_2 \in W_2 \}.$$

(a) Show that $W_1 \cap W_2$ and $W_1 + W_2$ are subspaces of V, and

$$W_1 \cap W_2 \subseteq W_1 \cup W_2 \subseteq W_1 + W_2.$$

(b) Explain the inclusions in (a) geometrically with two lines passing through the origin in the xy-plane.

(c) When is $W_1 \cup W_2$ a subspace of V?

(d) Show that $W_1 + W_2$ is the smallest subspace of V containing $W_1 \cup W_2$; that is, if S is a subspace of V containing $W_1 \cup W_2$, then $W_1 + W_2 \subseteq S$.

1.44 Let

$$W = \left\{ \begin{pmatrix} x_1 \\ x_2 \\ x_3 \\ x_4 \end{pmatrix} \in \mathbb{C}^4 \;\middle|\; x_3 = x_1 + x_2 \text{ and } x_4 = x_1 - x_2 \right\}.$$

(a) Prove that W is a subspace of $\mathbb{C}^4$.

(b) Find a basis for W. What is the dimension of W?

(c) Prove that $\{ k(1, 0, 1, 1)^t \mid k \in \mathbb{C} \}$ is a subspace of W.

1.45 Let V be a finite dimensional vector space and let V_1 and V_2 be subspaces of V. If $\dim(V_1 + V_2) = \dim(V_1 \cap V_2) + 1$, show that $V_1 + V_2$ is either V_1 or V_2 and $V_1 \cap V_2$ is correspondingly V_2 or V_1. Equivalently, for subspaces V_1 and V_2, if neither contains the other, then

$$\dim(V_1 + V_2) \geq \dim(V_1 \cap V_2) + 2.$$

1.46 Give an example of three subspaces of a vector space V such that

$$W_1 \cap (W_2 + W_3) \neq (W_1 \cap W_2) + (W_1 \cap W_3).$$

Why does this not contradict the following identity for any three sets

$$A \cap (B \cup C) = (A \cap B) \cup (A \cap C)?$$

1.47 Let W_1 and W_2 be nontrivial subspaces of a vector space V. The sum $W_1 + W_2$ is called a *direct sum*, denoted as $W_1 \oplus W_2$, if every element $\alpha \in W_1 + W_2$ can be uniquely written as $\alpha = w_1 + w_2$, where $w_1 \in W_1$ and $w_2 \in W_2$. Show that the following statements are equivalent:

(a) $W_1 + W_2$ is a direct sum.

(b) If $w_1 + w_2 = 0$ and $w_1 \in W_1$, $w_2 \in W_2$, then $w_1 = w_2 = 0$.

(c) $W_1 \cap W_2 = \{0\}$.

(d) $\dim(W_1 + W_2) = \dim W_1 + \dim W_2$.

How can the direct sum be extended to more than two subspaces?

1.48 Show that if W_1 is a subspace of a vector space V, and if there is a unique subspace W_2 such that $V = W_1 \oplus W_2$, then $W_1 = V$.

1.49 Let W_1, W_2, and W_3 be subspaces of a vector space V. Show that the sum $W_1 + W_2 + W_3 = \{ w_1 + w_2 + w_3 \mid w_i \in W_i,\ i = 1, 2, 3 \}$ is also a subspace of V. Show by example that $W_1 + W_2 + W_3$ is not necessarily a direct sum, i.e., there exist elements w_1, w_2, w_3, not all zero, such that $w_1 + w_2 + w_2 = 0$ and $w_i \in W_i$, $i = 1, 2, 3$, even though

$$W_1 \cap W_2 = W_1 \cap W_3 = W_2 \cap W_3 = \{0\}.$$

1.50 Show that $M_2(\mathbb{R}) = W_1 \oplus W_2$, where

$$W_1 = \left\{ \begin{pmatrix} a & b \\ -b & a \end{pmatrix} \,\middle|\, a,\ b \in \mathbb{R} \right\}$$

and

$$W_2 = \left\{ \begin{pmatrix} c & d \\ d & -c \end{pmatrix} \,\middle|\, c,\ d \in \mathbb{R} \right\}.$$

If V_1 is the subspace of $M_n(\mathbb{R})$ consisting of all $n \times n$ symmetric matrices, what will be a subspace V_2 such that $V_1 \oplus V_2 = M_n(\mathbb{R})$?

1.51 A function $f \in \mathcal{C}(\mathbb{R})$ is *even* if $f(-x) = f(x)$ for all $x \in \mathbb{R}$, and f is *odd* if $f(-x) = -f(x)$ for all $x \in \mathbb{R}$. Let W_1 and W_2 be the collections of even and odd continuous functions on $\mathbb{R}$, respectively. Show that W_1 and W_2 are subspaces of $\mathcal{C}(\mathbb{R})$. Show further that $\mathcal{C}(\mathbb{R}) = W_1 \oplus W_2$.

Chapter 2

Determinants, Inverses and Rank of Matrices, and Systems of Linear Equations

Definitions and Facts

Determinant. A *determinant* is a number assigned to a square matrix in a certain way. This number contains much information about the matrix. A very useful piece is that it tells immediately whether the matrix is invertible. For a square matrix A, we will denote its determinant by $|A|$ or $\det A$; both notations have been in common practice. Note that the bars are also used for the modulus of a complex number. However, one can usually tell from the context which use is intended.

If A is a 1×1 matrix; that is, A has one entry, say a_{11}, then its determinant is defined to be $|A| = a_{11}$. If A is a 2×2 matrix, say $A = \begin{pmatrix} a_{11} & a_{12} \\ a_{21} & a_{22} \end{pmatrix}$, then $|A|$ is given by $|A| = a_{11}a_{22} - a_{12}a_{21}$. The determinant for a square matrix with higher dimension n may be defined inductively as follows. Assume the determinant is defined for $(n-1) \times (n-1)$ matrices and let A_{1j} denote the submatrix of an $n \times n$ matrix A resulting from the deletion of the first row and the j-th column of the matrix A. Then

$$|A| = a_{11} \det A_{11} - a_{12} \det A_{12} + \cdots + (-1)^{1+n} a_{1n} \det A_{1n}.$$

The determinant can be defined in different, but equivalent, ways as follows: Let $A = (a_{ij})$ be an $n \times n$ matrix, $n \geq 2$, and let A_{ij} denote the $(n-1) \times (n-1)$ submatrix of A by deleting row i and column j from A,

$1 \leq i, j \leq n$. Then the determinant of A can be obtained by so-called *Laplace expansion* along row i; that is,

$$|A| = \sum_{j=1}^{n} (-1)^{i+j} a_{ij} |A_{ij}|.$$

It can be proved that the value $|A|$ is independent of choices of row i. Likewise, the Laplace expansion along a column may be defined.

The quantities $|A_{ij}|$ and $(-1)^{i+j}|A_{ij}|$ are called the *minor* and *cofactor* of the (i, j)-entry a_{ij}, respectively, and the matrix whose (i, j)-entry is the cofactor of a_{ji} is called the *adjoint* of the matrix A and denoted by $\operatorname{adj}(A)$. Let I be the n-square identity matrix. It follows that

$$A \operatorname{adj}(A) = |A|I.$$

Another definition of determinant in terms of permutations is concise and sometimes convenient. A permutation p on $\{1, 2, \ldots, n\}$ is said to be *even* if p can be restored to natural order by an even number of interchanges. Otherwise, p is *odd*. For instance, consider the permutations on $\{1, 2, 3, 4\}$. (There are $4! = 24$.) The permutation $p = (2, 1, 4, 3)$; that is, $p(1) = 2$, $p(2) = 1$, $p(3) = 4$, $p(4) = 3$, is even since it will become $(1, 2, 3, 4)$ after interchanging 2 and 1 and 4 and 3 (two interchanges), while $(1, 4, 3, 2)$ is odd, for interchanging 4 and 2 gives (1, 2, 3, 4).

Let S_n be the set of all permutations of $\{1, 2, \ldots, n\}$. For $p \in S_n$, define $\sigma(p) = +1$ if p is even and $\sigma(p) = -1$ if p is odd. It can be proved that

$$|A| = \sum_{p \in S_n} \sigma(p) \prod_{t=1}^{n} a_{tp(t)}.$$

Properties of Determinants. Let $A = (a_{ij})$ be an n-square matrix.

(d_1) A is singular if and only if $|A| = 0$.

(d_2) The rows (columns) of A are linearly dependent if and only if $|A| = 0$.

(d_3) If A is triangular, i.e., $a_{ij} = 0$, $i > j$ (or $i < j$), $|A| = a_{11}a_{22} \cdots a_{nn}$.

(d_4) $|A| = |A^t|$, where A^t is the transpose of A.

(d_5) $|kA| = k^n|A|$, where k is a scalar.

(d_6) $|AB| = |A|\,|B|$ for any n-square matrix B.

(d_7) $|S^{-1}AS| = |A|$ for any nonsingular n-square matrix S.

Elementary Row (Column) Operations on Matrices.

I. Interchange rows (columns) i and j.

II. Multiply row (column) i by a scalar $k \neq 0$.

III. Add k times row (column) i to row (column) j.

Suppose A is a square matrix, and B, C, and D are matrices obtained from A by the elementary row operations I, II, and III, respectively. Then

$$|B| = -|A|, \quad |C| = k|A|, \quad |D| = |A|.$$

Let E_{I}, E_{II}, and E_{III} denote the matrices obtained from the identity matrix by an application of I, II, and III, respectively, and call them *elementary matrices*. Then $B = E_{\mathrm{I}}A$, $C = E_{\mathrm{II}}A$, $D = E_{\mathrm{III}}A$. If an elementary column operation is applied to A, the resulting matrix is A postmultiplied by the corresponding elementary matrix.

Inverse. Let A be an $n \times n$ matrix. Matrix B is said to be an *inverse* of A if $AB = BA = I$. If A has an inverse, then its inverse is unique, and we denote it by A^{-1}. Moreover, since $|AA^{-1}| = |A|\,|A^{-1}| = 1$, it follows that

$$|A^{-1}| = \frac{1}{|A|}.$$

A square matrix A is *invertible* if and only if $|A| \neq 0$. In addition, if A is invertible, then so is its transpose A^t and $(A^t)^{-1} = (A^{-1})^t$; if A and B are invertible matrices of the same size, then AB is invertible and

$$(AB)^{-1} = B^{-1}A^{-1}.$$

Every invertible matrix is a product of some elementary matrices. This is seen by applications of a series of elementary row operations to the matrix to get the identity matrix I. When matrix A is invertible, the inverse can be found by the adjoint, the formula, however, is costly to calculate:

$$A^{-1} = \frac{1}{|A|}\operatorname{adj}(A).$$

For a 2×2 matrix, the following formula is convenient:

$$\text{If } A = \begin{pmatrix} a & b \\ c & d \end{pmatrix} \text{ and } ad - bc \neq 0, \text{ then } A^{-1} = \frac{1}{ad - bc}\begin{pmatrix} d & -b \\ -c & a \end{pmatrix}.$$

For a general matrix A with $|A| \neq 0$, one may find the inverse of A by converting the adjoined matrix (A, I) to a matrix in the form (I, B) through elementary row operations (reduction). Then B is A^{-1}.

Rank. Let A be an $m \times n$ matrix over a field $\mathbb{F}$, where $\mathbb{F} = \mathbb{R}$ or $\mathbb{C}$. The image or range of A, $\operatorname{Im} A = \{Ax \mid x \in \mathbb{F}^n\}$, is the column space of A and it is a subspace of $\mathbb{F}^m$ over $\mathbb{F}$. The *rank* of the matrix A is defined to be the dimension of its image (that is also a vector space over $\mathbb{F}$)

$$r(A) = \dim(\operatorname{Im} A).$$

Let A be an $m \times n$ matrix and P be an $n \times n$ invertible matrix. Then A and AP have the same column space, since, obviously, $\operatorname{Im}(AP) \subseteq \operatorname{Im} A$, and if $y = Ax \in \operatorname{Im} A$, then $y = Ax = (AP)(P^{-1}x) \in \operatorname{Im}(AP)$. It follows that applications of elementary column operations do not change the rank of a matrix. This is also true for row operations, because $\operatorname{Im} A$ and $\operatorname{Im}(QA)$ have the same dimension for any $m \times m$ invertible matrix Q. To see this, let $r(A) = r$ and take a basis $\alpha_1, \alpha_2, \dots, \alpha_r$ for $\operatorname{Im} A$, then $Q\alpha_1, Q\alpha_2, \dots, Q\alpha_r$ form a basis for $\operatorname{Im}(QA)$ and vice versa. Thus $\dim(\operatorname{Im} A) = \dim(\operatorname{Im}(QAP))$ for any $m \times m$ invertible matrix Q and any $n \times n$ invertible matrix P.

Let $A \neq 0$. Through elementary row and column operations, A can be brought to a matrix in the form $\begin{pmatrix} I_r & 0 \\ 0 & 0 \end{pmatrix}$; that is, there are invertible matrices R and S such that $A = R\begin{pmatrix} I_r & 0 \\ 0 & 0 \end{pmatrix} S$. In light of the above argument, we see that the rank of A is r. The following statements are true:

1. The dimension of the column (row) space of A is r; equivalently, the largest number of linearly independent columns (rows) of A is r.

2. There exists at least one $r \times r$ submatrix of A with nonzero determinant, and all $s \times s$ submatrices have zero determinant if $s > r$.

Other Properties of Rank. For matrices A, B, C of appreciate sizes,

(r_1) $r(A + B) \leq r(A) + r(B)$;

(r_2) $r(AB) \leq \min\{ r(A), r(B) \}$;

(r_3) $r(AB) + r(BC) - r(B) \leq r(ABC)$.

Systems of Linear Equations. Let $\mathbb{F}$ be a field and A be an $m \times n$ matrix over $\mathbb{F}$. Then $Ax = 0$ represents a *homogeneous* linear equation system of (m) linear equations (in n variables), where x is a column vector of n unknown components. The system $Ax = 0$ always has a solution $x = 0$.

If $r(A) = n$, then $x = 0$ is the unique solution. If $r(A) < n$, then $Ax = 0$ has infinitely many solutions. The solutions form a vector space, called the *solution space*, *null space*, or *kernel* of A, denoted by $\operatorname{Ker} A$.

Let $\{\alpha_1, \dots, \alpha_s\}$ be a basis for $\operatorname{Ker} A$ and extend it to a basis $\{\alpha_1, \dots, \alpha_s, \beta_1, \dots, \beta_t\}$ for $\mathbb{F}^n$, $s + t = n$. Obviously, $\operatorname{Im} A = \operatorname{Span}\{A\beta_1, \dots, A\beta_t\}$. If $A\beta_1, \dots, A\beta_t$ are linearly dependent, then for some $l_1, \dots, l_t$, not all zero, $l_1(A\beta_1) + \cdots + l_t(A\beta_t) = A(l_1\beta_1 + \cdots + l_t\beta_t) = 0$ and $l_1\beta_1 + \cdots + l_t\beta_t \in \operatorname{Ker} A$. This contradicts that $\{\alpha_1, \dots, \alpha_s, \beta_1, \dots, \beta_t\}$ is a basis for $\mathbb{F}^n$. Therefore $A\beta_1, \dots, A\beta_t$ are linearly independent and form a basis for $\operatorname{Im} A$. The dimension of the null space of A is $n - r(A)$; that is,

$$r(A) + \dim(\operatorname{Ker} A) = n.$$

Let b be a column vector of m components. Then the linear system $Ax = b$ may have one solution, infinitely many solutions, or no solution. These situations can be determined by the rank of the matrix $B = (A, b)$, which is obtained by augmenting b to A:

(s_1) If $r(B) = r(A) = n$, then $Ax = b$ has a unique solution.

(s_2) If $r(B) = r(A) < n$, then $Ax = b$ has infinitely many solutions.

(s_3) If $r(B) \neq r(A)$, then $Ax = b$ has no solution.

Cramer's Rule. Consider the linear equation system $Ax = b$, where A is a coefficient matrix of size $n \times n$. If A is invertible; that is, $|A| \neq 0$, then the system has a unique solution and the solution is given by

$$x_i = \frac{|A_i|}{|A|}, \qquad i = 1, 2, \dots, n,$$

where A_i is the matrix obtained from A by replacing the i-th column of A with b. Note that Cramer's rule cannot be used when A is singular.

Chapter 2 Problems

2.1 Evaluate the determinants

$$\begin{vmatrix} 1 & 2 & 3 \\ 8 & 9 & 4 \\ 7 & 6 & 5 \end{vmatrix}, \quad \begin{vmatrix} 1+x & 2+x & 3+x \\ 8+x & 9+x & 4+x \\ 7+x & 6+x & 5+x \end{vmatrix}, \quad \begin{vmatrix} x^1 & x^2 & x^3 \\ x^8 & x^9 & x^4 \\ x^7 & x^6 & x^5 \end{vmatrix}$$

2.2 Evaluate the determinant

$$\begin{vmatrix} 1 & 1 & 0 & 0 & 0 \\ -1 & 1 & 1 & 0 & 0 \\ 0 & -1 & 1 & 1 & 0 \\ 0 & 0 & -1 & 1 & 1 \\ 0 & 0 & 0 & -1 & 1 \end{vmatrix}$$

2.3 Explain without computation why the determinant equals zero:

$$\begin{vmatrix} a_1 & a_2 & a_3 & a_4 & a_5 \\ b_1 & b_2 & b_3 & b_4 & b_5 \\ c_1 & c_2 & 0 & 0 & 0 \\ d_1 & d_2 & 0 & 0 & 0 \\ e_1 & e_2 & 0 & 0 & 0 \end{vmatrix}$$

2.4 Evaluate the determinants

$$\begin{vmatrix} 0 & 0 & a_1 & b_1 \\ 0 & 0 & a_2 & b_2 \\ a_3 & b_3 & 0 & 0 \\ a_4 & b_4 & 0 & 0 \end{vmatrix}, \quad \begin{vmatrix} a_1 & 0 & 0 & b_1 \\ 0 & a_2 & b_2 & 0 \\ 0 & b_3 & a_3 & 0 \\ b_4 & 0 & 0 & a_4 \end{vmatrix}$$

2.5 Evaluate the 6×6 determinant

$$\begin{vmatrix} 0 & 0 & 0 & 0 & 0 & a_1 \\ 0 & 0 & 0 & 0 & a_2 & b \\ 0 & 0 & 0 & a_3 & c & d \\ 0 & 0 & a_4 & e & f & g \\ 0 & a_5 & h & i & j & k \\ a_6 & l & m & n & o & p \end{vmatrix}$$

2.6 Let $f(x) = (p_1 - x)(p_2 - x) \cdots (p_n - x)$ and let

$$\Delta_n = \begin{vmatrix} p_1 & a & a & a & \cdots & a & a \\ b & p_2 & a & a & \cdots & a & a \\ b & b & p_3 & a & \cdots & a & a \\ b & b & b & p_4 & \cdots & a & a \\ \vdots & \vdots & \vdots & \vdots & & \vdots & \vdots \\ b & b & b & b & \cdots & p_{n-1} & a \\ b & b & b & b & \cdots & b & p_n \end{vmatrix}$$

(a) Show that if $a \neq b$, then

$$\Delta_n = \frac{bf(a) - af(b)}{b - a}.$$

(b) Show that if $a = b$, then

$$\Delta_n = a \sum_{i=1}^{n} f_i(a) + p_n f_n(a),$$

where $f_i(a)$ means $f(a)$ with factor $(p_i - a)$ missing.

(c) Use (b) to evaluate

$$\begin{vmatrix} a & b & b & \cdots & b \\ b & a & b & \cdots & b \\ b & b & a & \cdots & a \\ \vdots & \vdots & \vdots & & \vdots \\ b & b & b & \cdots & a \end{vmatrix}_{n \times n}$$

2.7 Show that (the Vandermonde determinant)

$$\begin{vmatrix} 1 & 1 & 1 & \cdots & 1 \\ a_1 & a_2 & a_3 & \cdots & a_n \\ a_1^2 & a_2^2 & a_3^2 & \cdots & a_n^2 \\ \cdots & \cdots & \cdots & \cdots & \cdots \\ a_1^{n-1} & a_2^{n-1} & a_3^{n-1} & \cdots & a_n^{n-1} \end{vmatrix} = \prod_{1 \le i < j \le n} (a_j - a_i).$$

In particular, if V is the $n \times n$ matrix with (i, j)-entry j^{i-1}, then

$$|V| = (n-1)(n-2)^2 \cdots 2^{n-2}.$$

2.8 Show that if $a \neq b$, then

$$\begin{vmatrix} a+b & ab & 0 & \cdots & 0 & 0 \\ 1 & a+b & ab & \cdots & 0 & 0 \\ 0 & 1 & a+b & \cdots & 0 & 0 \\ \cdots & \cdots & \cdots & \cdots & \cdots & \cdots \\ 0 & 0 & 0 & \cdots & a+b & ab \\ 0 & 0 & 0 & \cdots & 1 & a+b \end{vmatrix}_{n\times n} = \frac{a^{n+1}-b^{n+1}}{a-b}.$$

What if $a = b$?

2.9 Find the characteristic polynomial $|\lambda I - A|$ for the 10×10 matrix

$$A = \begin{pmatrix} 0 & 1 & 0 & \ldots & 0 & 0 \\ 0 & 0 & 1 & \ldots & 0 & 0 \\ \cdots & \cdots & \cdots & \cdots & \cdots & \cdots \\ 0 & 0 & 0 & \ldots & 0 & 1 \\ 10^{10} & 0 & 0 & \ldots & 0 & 0 \end{pmatrix}.$$

2.10 Let $a_0, a_1, \ldots, a_{n-1} \in \mathbb{R}$. Write $a = (-a_1, -a_2, \ldots, -a_{n-1})$ and let

$$A = \begin{pmatrix} 0 & I_{n-1} \\ -a_0 & a \end{pmatrix}.$$

Show that

$$|\lambda I - A| = \lambda^n + a_{n-1}\lambda^{n-1} + \cdots + a_0.$$

2.11 Let each $a_{ij}(t)$ be a differentiable function of t. Show that

$$\frac{d}{dt}\begin{vmatrix} a_{11}(t) & \cdots & a_{1j}(t) & \cdots & a_{1n}(t) \\ a_{21}(t) & \cdots & a_{2j}(t) & \cdots & a_{2n}(t) \\ \cdots & \cdots & \cdots & \cdots & \cdots \\ a_{n1}(t) & \cdots & a_{nj}(t) & \cdots & a_{nn}(t) \end{vmatrix}$$

$$= \sum_{j=1}^{n} \begin{vmatrix} a_{11}(t) & \cdots & \frac{d}{dt}a_{1j}(t) & \cdots & a_{1n}(t) \\ a_{21}(t) & \cdots & \frac{d}{dt}a_{2j}(t) & \cdots & a_{2n}(t) \\ \cdots & \cdots & \cdots & \cdots & \cdots \\ a_{n1}(t) & \cdots & \frac{d}{dt}a_{nj}(t) & \cdots & a_{nn}(t) \end{vmatrix}$$

and evaluate $\frac{d}{dt}|I + tA|$ when $t = 0$.

2.12 If A is an $n \times n$ matrix all of whose entries are either 1 or -1, prove that $|A|$ is divisible by 2^{n-1}.

2.13 Let $A = (a,\ r_2,\ r_3,\ r_4)$ and $B = (b,\ r_2,\ r_3,\ r_4)$ be 4×4 matrices, where $a,\ b,\ r_2,\ r_3,\ r_4$ are column vectors in $\mathbb{R}^4$. If $\det A = 4$ and $\det B = 1$, find $\det(A+B)$. What is $\det C$, where $C = (r_4,\ r_3,\ r_2,\ a+b)$?

2.14 Let A be an $n \times n$ real matrix.

(a) Show that if $A^t = -A$ and n is odd, then $|A| = 0$.

(b) Show that if $A^2 + I = 0$, then n must be even.

(c) Does (b) remain true for complex matrices?

2.15 Let $A \in M_n(\mathbb{C})$. If $AA^t = I$ and $|A| < 0$, find $|A + I|$.

2.16 If A, B, C, D are $n \times n$ matrices such that $ABCD = I$. Show that

$$ABCD = DABC = CDAB = BCDA = I.$$

2.17 If A is such a matrix that $A^3 = 2I$, show that B is invertible, where

$$B = A^2 - 2A + 2I.$$

2.18 Consider $B(A, I) = (BA, B)$. On one hand, if B is the inverse of A, then (BA, B) becomes (I, A^{-1}). On the other hand, B is a product of elementary matrices since it is invertible. This indicates that the inverse of A can be obtained by applying elementary row operations to the augmented matrix (A, I) to get (I, A^{-1}). Find the inverses of

$$A = \begin{pmatrix} 1 & 0 & 0 \\ 0 & 1 & 0 \\ a & b & 1 \end{pmatrix} \quad \text{and} \quad B = \begin{pmatrix} 0 & 1 & 0 & 0 \\ 0 & 0 & 1 & 0 \\ 0 & 0 & 0 & 1 \\ a & b & c & d \end{pmatrix}.$$

2.19 Find the inverses of the matrices

$$\begin{pmatrix} 1 & 1 & 1 \\ 0 & 1 & 1 \\ 0 & 0 & 1 \end{pmatrix} \quad \text{and} \quad \begin{pmatrix} 1 & 1 & \dots & 1 & 1 \\ 0 & 1 & \dots & 1 & 1 \\ \vdots & \vdots & \ddots & \vdots & \vdots \\ 0 & 0 & \dots & 1 & 1 \\ 0 & 0 & \dots & 0 & 1 \end{pmatrix}.$$

2.20 Find the inverse of the matrix

$$\begin{pmatrix} 1 & 1 & 1 & \dots & 1 \\ 1 & 2 & 2 & \dots & 2 \\ 1 & 2 & 3 & \dots & 3 \\ \vdots & \vdots & \vdots & \ddots & \vdots \\ 1 & 2 & 3 & \dots & n \end{pmatrix}.$$

2.21 Find the determinant and inverse of the $n \times n$ matrix

$$\begin{pmatrix} 0 & 1 & 1 & \dots & 1 \\ 1 & 0 & 1 & \dots & 1 \\ 1 & 1 & 0 & \dots & 1 \\ \vdots & \vdots & \vdots & \ddots & \vdots \\ 1 & 1 & 1 & \dots & 0 \end{pmatrix}.$$

2.22 Let $a_1, a_2, \dots, a_n$ be nonzero numbers. Find the inverse of the matrix

$$\begin{pmatrix} 0 & a_1 & 0 & \dots & 0 \\ 0 & 0 & a_2 & \dots & 0 \\ \vdots & \vdots & \ddots & \ddots & \vdots \\ 0 & 0 & 0 & \dots & a_{n-1} \\ a_n & 0 & 0 & \dots & 0 \end{pmatrix}.$$

2.23 Let A, B, C, X, Y, $Z \in M_n(\mathbb{C})$, and A^{-1} and C^{-1} exist. Find

$$\begin{pmatrix} A & B \\ 0 & C \end{pmatrix}^{-1} \quad \text{and} \quad \begin{pmatrix} I & X & Y \\ 0 & I & Z \\ 0 & 0 & I \end{pmatrix}^{-1}.$$

2.24 Find the inverse of the 3×3 Vandermonde matrix

$$V = \begin{pmatrix} 1 & 1 & 1 \\ a_1 & a_2 & a_3 \\ a_1^2 & a_2^2 & a_3^2 \end{pmatrix}$$

when a_1, a_2, and a_3 are distinct from each other.

2.25 Let A and B be, respectively, $m \times p$ and $m \times q$ matrices such that $A^*B = 0$, where $p + q = m$. If $M = (A, B)$ is invertible, show that

$$M^{-1} = \begin{pmatrix} (A^*A)^{-1}A^* \\ (B^*B)^{-1}B^* \end{pmatrix}.$$

2.26 Assuming that all matrix inverses involved below exist, show that

$$(A - B)^{-1} = A^{-1} + A^{-1}(B^{-1} - A^{-1})^{-1}A^{-1}.$$

In particular

$$(I + A)^{-1} = I - (A^{-1} + I)^{-1}$$

and

$$|(I + A)^{-1} + (I + A^{-1})^{-1}| = 1.$$

2.27 Assuming that all matrix inverses involved below exist, show that

$$(A + iB)^{-1} = B^{-1}A(A + AB^{-1}A)^{-1} - i(B + AB^{-1}A)^{-1}.$$

2.28 Let A, B, C, $D \in M_n(\mathbb{C})$. If AB and CD are Hermitian, show that

$$AD - B^*C^* = I \quad \Rightarrow \quad DA - BC = I.$$

2.29 Let m and n be positive integers and denote $K = \left(\begin{smallmatrix} I_m & 0 \\ 0 & -I_n \end{smallmatrix}\right)$. Let S_K be the collection of all $(m+n)$-square complex matrices X such that

$$X^*KX = K.$$

(a) If $A \in S_K$, show that A^{-1} exists and A^{-1}, A^t, $\bar{A}$, $A^* \in S_K$.

(b) If A, $B \in S_K$, show that $AB \in S_K$. How about kA or $A + B$?

(c) Discuss a similar problem with $K = \left(\begin{smallmatrix} 0 & I_m \\ -I_m & 0 \end{smallmatrix}\right)$.

2.30 Let A and C be $m \times m$ and $n \times n$ matrices, respectively, and let B, D, and E be matrices of appropriate sizes.

(a) Show that

$$\begin{vmatrix} A & B \\ 0 & C \end{vmatrix} = \begin{vmatrix} A & 0 \\ D & C \end{vmatrix} = |A||C|.$$

(b) Evaluate

$$\begin{vmatrix} I_m & 0 \\ 0 & I_n \end{vmatrix}, \quad \begin{vmatrix} 0 & I_m \\ I_n & 0 \end{vmatrix}, \quad \begin{vmatrix} I_m & B \\ 0 & I_n \end{vmatrix}.$$

(c) Find a formula for

$$\begin{vmatrix} 0 & A \\ C & E \end{vmatrix}.$$

2.31 Let S be the *backward identity* matrix; that is,

$$S = \begin{pmatrix} 0 & 0 & \cdots & 0 & 1 \\ 0 & 0 & \cdots & 1 & 0 \\ \vdots & \vdots & & \vdots & \vdots \\ 0 & 1 & \cdots & 0 & 0 \\ 1 & 0 & \cdots & 0 & 0 \end{pmatrix}_{n \times n}.$$

Show that $S^{-1} = S^t = S$. Find $|S|$ and SAS for $A = (a_{ij}) \in M_n(\mathbb{C})$.

2.32 Let A, B, C, D be $m \times p$, $m \times q$, $n \times p$, $n \times q$ matrices, respectively, where $m + n = p + q$. Show that

$$\begin{vmatrix} A & B \\ C & D \end{vmatrix} = (-1)^{(mn+pq)} \begin{vmatrix} D & C \\ B & A \end{vmatrix}.$$

In particular, when A, B, C, D are square matrices of the same size,

$$\begin{vmatrix} A & B \\ C & D \end{vmatrix} = \begin{vmatrix} D & C \\ B & A \end{vmatrix},$$

and for a square matrix A, a column vector x, and a row vector y,

$$\begin{vmatrix} A & x \\ y & 1 \end{vmatrix} = \begin{vmatrix} 1 & y \\ x & A \end{vmatrix}.$$

Is it true in general that

$$\begin{vmatrix} A & B \\ C & D \end{vmatrix} = \begin{vmatrix} A & C \\ B & D \end{vmatrix}?$$

2.33 Let A, B, C, $D \in M_n(\mathbb{C})$. If matrix $\left(\begin{smallmatrix} A & B \\ C & D \end{smallmatrix}\right)$ has rank n, show that

$$\begin{vmatrix} |A| & |B| \\ |C| & |D| \end{vmatrix} = 0.$$

Moreover, if A is invertible, then $D = CA^{-1}B$.

2.34 Let A, B, C, $D \in M_n(\mathbb{C})$ and let $M = \left(\begin{smallmatrix} A & B \\ C & D \end{smallmatrix}\right)$. Show that

(a) $|M| = |AD^t - BC^t|$ if $CD^t = DC^t$.

(b) $|M| = |AD^t + BC^t|$ if $CD^t + DC^t = 0$ and if D^{-1} exists.

(c) (b) is invalid if D is singular by example.

(d) $|M|^2 = |AD^t + BC^t|^2$ for the example constructed in (c).

2.35 Let A, B, C, $D \in M_n(\mathbb{C})$.

(a) Show that if A^{-1} exists, then

$$\begin{vmatrix} A & B \\ C & D \end{vmatrix} = |A|\,|D - CA^{-1}B|.$$

(b) Show that if $AC = CA$, then

$$\begin{vmatrix} A & B \\ C & D \end{vmatrix} = |AD - CB|.$$

(c) Can B and C on the right-hand side in (b) be switched?

(d) Does (b) remain true if the condition $AC = CA$ is dropped?

2.36 Consider the matrices in $M_2(\mathbb{R})$.

(a) Is it true that $|A + B| = |A| + |B|$ in general?

(b) If $A \neq 0$, B_1, B_2, B_3, $B_4 \in M_2(\mathbb{R})$, and if

$$|A + B_i| = |A| + |B_i|, \quad i = 1, 2, 3, 4,$$

show that B_1, B_2, B_3, B_4 are linearly dependent over $\mathbb{R}$.

2.37 Let $M = \begin{pmatrix} A & B \\ C & D \end{pmatrix}$ be an invertible matrix with $M^{-1} = \begin{pmatrix} X & U \\ Y & V \end{pmatrix}$, where A and D are square matrices (possibly of different sizes), B and C are matrices of appropriate sizes, and X has the same size as A.

(a) Show that $|A| = |V|\,|M|$.

(b) If A is invertible, show that $X = (D - CA^{-1}B)^{-1}$.

(c) Consider a unitary matrix W partitioned as $W = \begin{pmatrix} u & x \\ y & U_1 \end{pmatrix}$, where u is a number and U_1 is a square matrix. Show that u and $\det U_1$ have the same modulus; that is, $|u| = |\det U_1|$.

(d) What conclusion can be drawn for real orthogonal matrices?

2.38 Introduce the following correspondences between complex numbers and real matrices and between complex number pairs and complex matrices:

$$z = x + iy \sim Z = \begin{pmatrix} x & y \\ -y & x \end{pmatrix} \in M_2(\mathbb{R}),$$

$$q = (u, v) \simeq Q = \begin{pmatrix} u & v \\ -\bar{v} & \bar{u} \end{pmatrix} \in M_2(\mathbb{C}).$$

(a) Show that $\bar{z} \sim Z^t$.

(b) Show that $ZW = WZ$, where $w \sim W$.

(c) Show that $z \sim Z$ and $w \sim W$ imply $zw \sim ZW$.

(d) Find Z^n for $z = r(\cos\theta + i\sin\theta)$, r, $\theta \in \mathbb{R}$.

(e) What is the matrix corresponding to $z = i$?

(f) Show that $Z^{-1} = \frac{1}{x^2+y^2}\begin{pmatrix} x & -y \\ y & x \end{pmatrix}$.

(g) Show that $Z = P\begin{pmatrix} x+iy & 0 \\ 0 & x-iy \end{pmatrix}P^*$, where $P = \frac{1}{\sqrt{2}}\begin{pmatrix} 1 & 1 \\ i & -i \end{pmatrix}$.

(h) Show that $|Q| \geq 0$. Find Q^{-1} when $|u|^2 + |v|^2 = 1$.

(i) Replace each entry z of Q with the corresponding 2×2 real matrix Z to get $R = \begin{pmatrix} U & V \\ -V^t & U \end{pmatrix} \in M_4(\mathbb{R})$. Show that $|R| \geq 0$.

(j) Show that R in (i) is similar to a matrix of the form $\begin{pmatrix} U & X \\ -X & U \end{pmatrix}$.

(k) Show that R in (i) is singular if and only if Q is singular, and if and only if $u = v = 0$.

2.39 Let A and B be $n \times n$ real matrices. Show that

$$\begin{vmatrix} A & B \\ -B & A \end{vmatrix} \geq 0.$$

State the analog for complex matrices.

2.40 Let A and B be $n \times n$ complex matrices. Show that

$$\begin{vmatrix} A & B \\ B & A \end{vmatrix} = |A+B||A-B|.$$

Let $C = A + B$ and $D = A - B$. If C and D are invertible, show that

$$\begin{pmatrix} A & B \\ B & A \end{pmatrix}^{-1} = \frac{1}{2}\begin{pmatrix} C^{-1}+D^{-1} & C^{-1}-D^{-1} \\ C^{-1}-D^{-1} & C^{-1}+D^{-1} \end{pmatrix}.$$

2.41 Let x and y be column vectors of n complex components. Show that

(a) $|I - xy^*| = 1 - y^*x$.

(b) $\begin{vmatrix} I & x \\ y^* & 1 \end{vmatrix} = \begin{vmatrix} 1 & y^* \\ x & I \end{vmatrix}$.

(c) If $\delta = 1 - y^*x \neq 0$, then $(I - xy^*)^{-1} = I + \delta^{-1}xy^*$.

(d) $\begin{pmatrix} I & x \\ y^* & 1 \end{pmatrix}^{-1} = \begin{pmatrix} I + \delta^{-1}xy^* & -\delta^{-1}x \\ -\delta^{-1}y^* & \delta^{-1} \end{pmatrix}$.

2.42 Show that a matrix A is of rank 1 if and only if A can be written as $A = xy^t$ for some column vectors x and y.

2.43 Let $A \in M_n(\mathbb{C})$ and $u_1, u_2, \ldots, u_n \in \mathbb{C}^n$ be linearly independent. Show that $r(A) = n$, namely A is nonsingular, if and only if $Au_1, Au_2, \ldots, Au_n$ are linearly independent.

2.44 Let $A \neq 0$ be an $m \times n$ complex matrix with rank r, show that there exist invertible $m \times m$ matrix P and $n \times n$ matrix Q such that

$$A = P\begin{pmatrix} I_r & 0 \\ 0 & 0 \end{pmatrix}Q.$$

Moreover, P and Q can be chosen to be real if A is real.

2.45 For matrices of appropriate sizes, answer true or false:

(a) If $A^2 = B^2$, then $A = B$ or $A = -B$.

(b) If $r(A) = r(B)$, then $r(A^2) = r(B^2)$.

(c) $r(A + kB) \leq r(A) + kr(B)$, where k is a positive scalar.

(d) $r(A - B) \leq r(A) - r(B)$.

(e) If $r(AB) = 0$, then $r(BA) = 0$.

(f) If $r(AB) = 0$, then $r(A) = 0$ or $r(B) = 0$.

2.46 Consider the 2×2 Hermitian matrix $A = \begin{pmatrix} 1 & i \\ -i & 1 \end{pmatrix}$. Let

$$W_{\mathbb{R}} = \{ Ax \mid x \in \mathbb{R}^2 \} \quad \text{and} \quad W_{\mathbb{C}} = \{ Ax \mid x \in \mathbb{C}^2 \}.$$

(a) Show that the rows of A are linearly dependent over $\mathbb{C}$.

(b) Show that the rows of A are linearly independent over $\mathbb{R}$.

(c) Since the rows of A are linearly independent over the real number field $\mathbb{R}$, does it follow that the matrix A is invertible?

(d) Show that U^*AU is a diagonal matrix, where $U = \frac{1}{\sqrt{2}} \begin{pmatrix} i & -i \\ 1 & 1 \end{pmatrix}$.

(e) What is the rank of A?

(f) Show that $W_{\mathbb{R}} \subseteq W_{\mathbb{C}}$.

(g) Show that $W_{\mathbb{C}}$ is a subspace of $\mathbb{C}^2$ over $\mathbb{R}$ and also over $\mathbb{C}$.

(h) Show that $W_{\mathbb{R}}$ is a subspace of $\mathbb{C}^2$ over $\mathbb{R}$ but not over $\mathbb{C}$.

(i) Find $\dim W_{\mathbb{R}}$ over $\mathbb{R}$ and $\dim W_{\mathbb{C}}$ over $\mathbb{R}$ and over $\mathbb{C}$.

2.47 Let A be an n-square Hermitian matrix. Write $A = B + iC$, where B and C are n-square real matrices.

(a) Show that $B^t = B$ and $C^t = -C$.

(b) Show that $x^tAx = x^tBx$ and $x^tCx = 0$ for all $x \in \mathbb{R}^n$.

(c) Show that if $Ax = 0$, $x \in \mathbb{R}^n$, then $Bx = 0$ and $Cx = 0$.

(d) Take $A = \begin{pmatrix} 1 & i \\ -i & 1 \end{pmatrix}$. Find a complex column vector $x \in \mathbb{C}^2$ such that $x^*Ax = 0$ but $x^*Bx \neq 0$.

(e) Take $A = \begin{pmatrix} 1 & 1+i \\ 1-i & 1 \end{pmatrix}$. Find a real column vector $x \in \mathbb{R}^2$ such that $Bx = 0$ but $Ax \neq 0$. What are the ranks of A and B?

2.48 For an $m \times n$ real matrix A, let $W_{\mathbb{R}} = \{ Ax \mid x \in \mathbb{R}^n \}$ and $W_{\mathbb{C}} = \{Ax \mid x \in \mathbb{C}^n \}$. Then obviously $W_{\mathbb{R}}$ and $W_{\mathbb{C}}$ are not the same. Show that the dimension of $W_{\mathbb{R}}$ as a subspace of $\mathbb{R}^m$ over $\mathbb{R}$ is the same as the dimension of $W_{\mathbb{C}}$ as a subspace of $\mathbb{C}^m$ over $\mathbb{C}$.

2.49 If the rank of the following 3×4 matrix A is 2, find the value of t:

$$A = \begin{pmatrix} 1 & 2 & -1 & 1 \\ 2 & 0 & t & 0 \\ 0 & -4 & 5 & 2 \end{pmatrix}.$$

2.50 For what value of t is the rank of the following matrix A equal to 3?

$$A = \begin{pmatrix} t & 1 & 1 & 1 \\ 1 & t & 1 & 1 \\ 1 & 1 & t & 1 \\ 1 & 1 & 1 & t \end{pmatrix}.$$

2.51 Let $A, B \in M_n(\mathbb{C})$. Show that if $AB = 0$, then

$$r(A) + r(B) \leq n.$$

2.52 If B is a submatrix of a matrix A obtained by deleting s rows and t columns from A, show that

$$r(A) \leq s + t + r(B).$$

2.53 Let $A, B \in M_n(\mathbb{C})$. Show that

$$r(AB) \leq \min\{ r(A), r(B) \}$$

and

$$r(A + B) \leq r(A) + r(B) \leq r(AB) + n.$$

2.54 Let A be $m \times n$, B be $n \times p$, and C be $p \times q$ matrices. Show that

$$r(ABC) \geq r(AB) + r(BC) - r(B).$$

2.55 Let $A, B \in M_n(\mathbb{C})$. If $AB = BA$, show that

$$r(A + B) \leq r(A) + r(B) - r(AB).$$

2.56 Let $A_1, A_2, \ldots, A_k$ be $n \times n$ matrices. If $A_1 A_2 \cdots A_k = 0$, show that

$$r(A_1) + r(A_2) + \cdots + r(A_k) \le (k-1)n.$$

2.57 Let X, Y, and Z be matrices of the same number of rows. Show that

$$r(X,Y) \le r(X,Z) + r(Z,Y) - r(Z).$$

2.58 Show that for any $m \times n$ complex matrix A,

$$r(A^*A) = r(AA^*) = r(A) = r(A^*) = r(A^t) = r(\bar{A}).$$

Is $r(A^tA)$ or $r(\bar{A}A)$ equal to $r(A)$ in general?

2.59 Which of the following is M^* for the partitioned matrix $M = \begin{pmatrix} A & B \\ C & D \end{pmatrix}$?

$$M_1 = \begin{pmatrix} A^* & B^* \\ C^* & D^* \end{pmatrix}, \quad M_2 = \begin{pmatrix} A^* & C^* \\ B^* & D^* \end{pmatrix}, \quad M_3 = \begin{pmatrix} A & B^* \\ C^* & D \end{pmatrix}, \quad M_4 = \begin{pmatrix} \bar{A} & C^* \\ B^* & \bar{D} \end{pmatrix}.$$

2.60 Let $\operatorname{adj}(A)$ denote the *adjoint* of $A \in M_n(\mathbb{C})$; that is, $\operatorname{adj}(A)$ is the $n \times n$ matrix whose (i,j)-entry is the cofactor $(-1)^{i+j}|A_{ji}|$ of a_{ji}, where A_{ji} is the submatrix obtained from A by deleting the j-th row and the i-th column. Show that

(a) $r(A) = n$ if and only if $r(\operatorname{adj}(A)) = n$.

(b) $r(A) = n-1$ if and only if $r(\operatorname{adj}(A)) = 1$.

(c) $r(A) < n-1$ if and only if $r(\operatorname{adj}(A)) = 0$.

(d) $|\operatorname{adj}(A)| = |A|^{n-1}$.

(e) $\operatorname{adj}(\operatorname{adj}(A)) = |A|^{n-2}A$.

(f) $\operatorname{adj}(AB) = \operatorname{adj}(B)\operatorname{adj}(A)$.

(g) $\operatorname{adj}(XAX^{-1}) = X(\operatorname{adj}(A))X^{-1}$ for any invertible $X \in M_n(\mathbb{C})$.

(h) $|\overbrace{\operatorname{adj}\cdots\operatorname{adj}}^{k}(A)| = |A|$ when A is 2×2.

(i) If A is Hermitian, so is $\operatorname{adj}(A)$.

Find a formula for $\overbrace{\operatorname{adj}\cdots\operatorname{adj}}^{k}(A)$ when $|A| = 1$. What are the eigenvalues of $\operatorname{adj}(A)$? What are the eigenvalues of $\overbrace{\operatorname{adj}\cdots\operatorname{adj}}^{k}(A)$?

2.61 Show that A is nonsingular if $A = (a_{ij}) \in M_n(\mathbb{C})$ satisfies

$$|a_{ii}| > \sum_{j=1,\ j\neq i}^{n} |a_{ij}|, \quad i = 1, 2, \ldots, n.$$

2.62 Let A, B be $n \times n$ matrices satisfying $A^2 = A$, $B^2 = B$. Show that

$$r(A - B) = r(A - AB) + r(B - AB).$$

2.63 Let $A \in M_n(\mathbb{C})$. Show that $A^2 = A$ if and only if $r(A) + r(A - I) = n$.

2.64 Let A be an $m \times n$ matrix and B be an $n \times p$ matrix. Show that

$$r\begin{pmatrix} 0 & A \\ B & I_n \end{pmatrix} = n + r(AB).$$

2.65 Let A be an $m \times n$ matrix, $m \geq n$. Show that

$$r(I_m - AA^*) - r(I_n - A^*A) = m - n.$$

2.66 Denote the columns of matrix A by α_1, α_2, α_3, respectively, where

$$A = \begin{pmatrix} 1+\lambda & 1 & 1 \\ 1 & 1+\lambda & 1 \\ 1 & 1 & 1+\lambda \end{pmatrix}.$$

Find the value(s) of λ such that $\beta = (0,\ \lambda,\ \lambda^2)^t$

(a) belongs to the column space of A;

(b) does not belong to the column space of A.

2.67 The notation A^* is used for the adjoint of matrix A in many other books. Under what conditions on the matrix A is A^* in the sense of this book; that is, $A^* = (\bar{A})^t$, the conjugate transpose, the same as $\operatorname{adj}(A)$, the adjoint matrix of A?

2.68 Determine the values of λ so that the following linear equation system of three unknowns has only the zero solution:

$$\begin{aligned} \lambda x_1 + x_2 + x_3 &= 0 \\ x_1 + \lambda x_2 + x_3 &= 0 \\ x_1 + x_2 + x_3 &= 0. \end{aligned}$$

2.69 Determine the value of λ so that the following linear equation system of three unknowns has nonzero solutions:

$$\begin{aligned} x_1 + 2x_2 - 2x_3 &= 0 \\ 2x_1 - x_2 + \lambda x_3 &= 0 \\ 3x_1 + x_2 - x_3 &= 0. \end{aligned}$$

2.70 Find the general solutions of the linear system of five unknowns:

$$\begin{aligned} x_1 + x_2 + x_5 &= 0 \\ x_1 + x_2 - x_3 &= 0 \\ x_3 + x_4 + x_5 &= 0. \end{aligned}$$

2.71 Find the dimension and a basis for the solution space of the system

$$\begin{aligned} x_1 - x_2 + 5x_3 - x_4 &= 0 \\ x_1 + x_2 - 2x_3 + 3x_4 &= 0 \\ 3x_1 - x_2 + 8x_3 + x_4 &= 0 \\ x_1 + 3x_2 - 9x_3 + 7x_4 &= 0. \end{aligned}$$

2.72 Find all solutions x_1, x_2, x_3, x_4, x_5 of the linear equation system

$$\begin{aligned} x_5 + x_2 &= yx_1 \\ x_1 + x_3 &= yx_2 \\ x_2 + x_4 &= yx_3 \\ x_3 + x_5 &= yx_4 \\ x_4 + x_1 &= yx_5, \end{aligned}$$

where y is a parameter.

2.73 Discuss the solutions of the equation system in unknowns x_1, x_2, x_3:

$$\begin{aligned} ax_1 + bx_2 + 2x_3 &= 1 \\ ax_1 + (2b-1)x_2 + 3x_3 &= 1 \\ ax_1 + bx_2 + (b+3)x_3 &= 2b-1. \end{aligned}$$

2.74 Let A be a real matrix. If the linear equation system $Ax = 0$ has a nonzero complex solution, show that it has a nonzero real solution.

2.75 Find a basis for the solution space of the system of $n+1$ linear equations of $2n$ unknowns:

$$\begin{aligned} x_1 + x_2 + \cdots + x_n &= 0 \\ x_2 + x_3 + \cdots + x_{n+1} &= 0 \\ &\vdots \\ x_{n+1} + x_{n+2} + \cdots + x_{2n} &= 0. \end{aligned}$$

2.76 Let $A \in M_n(\mathbb{F})$ and write $A = (\alpha_1, \alpha_2, \ldots, \alpha_n)$, where each $\alpha_i \in \mathbb{F}^n$.

(a) Show that

$$\dim(\operatorname{Span}\{\alpha_1, \alpha_2, \ldots, \alpha_n\}) = r(A).$$

(b) Let P be an $n \times n$ invertible matrix. Write

$$PA = (P\alpha_1, P\alpha_2, \ldots, P\alpha_n) = (\beta_1, \beta_2, \ldots, \beta_n).$$

Show that $\alpha_{i_1}, \alpha_{i_2}, \ldots, \alpha_{i_r}$ are linearly independent if and only if $\beta_{i_1}, \beta_{i_2}, \ldots, \beta_{i_r}$ are linearly independent (over $\mathbb{F}$).

(c) Find the dimension and a basis of the space spanned by

$$\gamma_1 = \begin{pmatrix} 2 \\ 1 \\ 3 \\ 1 \end{pmatrix}, \quad \gamma_2 = \begin{pmatrix} 1 \\ 2 \\ 0 \\ 1 \end{pmatrix}, \quad \gamma_3 = \begin{pmatrix} 0 \\ 2 \\ -2 \\ 1 \end{pmatrix}, \quad \gamma_4 = \begin{pmatrix} 1 \\ 1 \\ 1 \\ 1 \end{pmatrix}.$$

2.77 Let W_1 and W_2 be the vector spaces over $\mathbb{R}$ spanned, respectively, by

$$\alpha_1 = \begin{pmatrix} 1 \\ 2 \\ -1 \\ -2 \end{pmatrix}, \quad \alpha_2 = \begin{pmatrix} 3 \\ 1 \\ 1 \\ 1 \end{pmatrix}, \quad \alpha_3 = \begin{pmatrix} -1 \\ 0 \\ 1 \\ -1 \end{pmatrix}$$

and

$$\beta_1 = \begin{pmatrix} 2 \\ 5 \\ -6 \\ -5 \end{pmatrix}, \quad \beta_2 = \begin{pmatrix} -1 \\ 2 \\ -7 \\ 3 \end{pmatrix}.$$

Find the dimensions and bases for $W_1 \cap W_2$ and $W_1 + W_2$.

2.78 Let a_{ij} be integers, $1 \leq i, j \leq n$. If for any set of integers $b_1, b_2, \ldots, b_n$, the system of linear equations

$$\sum_{j=1}^{n} a_{ij}x_j = b_j, \quad i = 1, 2, \ldots, n,$$

has integer solutions $x_1, x_2, \ldots, x_n$, show that the determinant of the coefficient matrix $A = (a_{ij})$ is either 1 or -1.

2.79 Let W_1 and W_2 be the solution spaces of the linear equation systems

$$x_1 + x_2 + \cdots + x_n = 0$$

and

$$x_1 = x_2 = \cdots = x_n,$$

respectively, where $x_i \in \mathbb{F}$, $i = 1, 2, \ldots, n$. Show that $\mathbb{F}^n = W_1 \oplus W_2$.

2.80 Let $A \in M_n(\mathbb{C})$. Show that there exists an $n \times n$ nonzero matrix B such that $AB = 0$ if and only if $|A| = 0$.

2.81 Let A be a $p \times n$ matrix and B be a $q \times n$ matrix over a field $\mathbb{F}$. If $r(A) + r(B) < n$, show that there must exist a nonzero column vector x of n components such that both $Ax = 0$ and $Bx = 0$.

2.82 Let A and B be n-square matrices over a field $\mathbb{F}$ and $\operatorname{Ker} A$ and $\operatorname{Ker} B$ be the null spaces of A and B with dimensions l and m, respectively. Show that the null space of AB has dimension at least $\max\{l, m\}$. When does it happen that every $x \in \mathbb{F}^n$ is either in $\operatorname{Ker} A$ or $\operatorname{Ker} B$?

2.83 Let A be a square matrix. If $r(A) = r(A^2)$, show that the equation systems $Ax = 0$ and $A^2x = 0$ have the same solution space.

2.84 Let A and B be $m \times n$ matrices. Show that $Ax = 0$ and $Bx = 0$ have the same solution space if and only if there exists an invertible matrix C such that $A = CB$. Use this fact to show that if $r(A^2) = r(A)$, then there exists an invertible matrix D such that $A^2 = DA$.

2.85 Suppose $b \neq 0$ and $Ax = b$ has solutions $\eta_1, \eta_2, \dots, \eta_n$. Show that a linear combination $\lambda_1\eta_1 + \lambda_2\eta_2 + \cdots + \lambda_n\eta_n$ is a solution to $Ax = b$ if and only if $\lambda_1 + \lambda_2 + \cdots + \lambda_n = 1$. Show also that $l_1\eta_1 + l_2\eta_2 + \cdots + l_n\eta_n = 0$ implies $l_1 + l_2 + \cdots + l_n = 0$.

2.86 Let $A \in M_{m\times n}(\mathbb{C})$. Show that for any $b \in \mathbb{C}^m$, the linear equation system $A^*Ax = A^*b$ is consistent, meaning that it has solutions.

2.87 Let $A \in M_n(\mathbb{C})$ and b be a column vector of n complex complements. Denote $\tilde{A} = \begin{pmatrix} A & b \\ b^* & 0 \end{pmatrix}$. If $r(\tilde{A}) = r(A)$, which of the following is true?

(a) $Ax = b$ has infinitely many solutions.

(b) $Ax = b$ has a unique solution.

(c) $\tilde{A}x = 0$ has only solution $x = 0$.

(d) $\tilde{A}x = 0$ has nonzero solutions.

2.88 Let

$$\begin{vmatrix} a_{11} & a_{12} & \dots & a_{1n} \\ a_{21} & a_{22} & \dots & a_{2n} \\ \dots & \dots & \dots & \dots \\ a_{n1} & a_{n2} & \dots & a_{nn} \end{vmatrix} \neq 0.$$

Show that

$$\begin{cases} a_{11}x_1 + a_{12}x_2 + \cdots + a_{1n}x_n = b_1 \\ a_{21}x_1 + a_{22}x_2 + \cdots + a_{2n}x_n = b_2 \\ \dots\dots\dots\dots\dots\dots\dots\dots\dots\dots \\ a_{n1}x_1 + a_{n2}x_2 + \cdots + a_{nn}x_n = b_n \\ c_1x_1 + c_2x_2 + \cdots + c_nx_n = d \end{cases}$$

and

$$\begin{cases} a_{11}x_1 + a_{21}x_2 + \cdots + a_{n1}x_n = c_1 \\ a_{12}x_1 + a_{22}x_2 + \cdots + a_{n2}x_n = c_2 \\ \dots\dots\dots\dots\dots\dots\dots\dots\dots\dots \\ a_{1n}x_1 + a_{2n}x_2 + \cdots + a_{nn}x_n = c_n \\ b_1x_1 + b_2x_2 + \cdots + b_nx_n = d \end{cases}$$

will either both have a unique solution or both have no solution.

2.89 If A is a square matrix such that the linear equation system $Ax = 0$ has nonzero solutions, is it possible that $A^t x = b$ has a unique solution for some column vector b?

2.90 Let A and B be matrices such that $r(AB) = r(A)$. Show that

$$X_1 AB = X_2 AB \quad \Rightarrow \quad X_1 A = X_2 A.$$

2.91 Consider the straight lines in the xy-plane. Show that the three lines

$$\begin{aligned} l_1 : &\quad ax + by + c = 0 \\ l_2 : &\quad bx + cy + a = 0 \\ l_3 : &\quad cx + ay + b = 0 \end{aligned}$$

intersect at a point if and only if $a + b + c = 0$.

Chapter 3

Matrix Similarity, Eigenvalues, Eigenvectors, and Linear Transformations

Definitions and Facts

Similarity. Let A and B be $n \times n$ matrices over a field $\mathbb{F}$. If there exists an $n \times n$ invertible matrix P over $\mathbb{F}$ such that $P^{-1}AP = B$, we say that A and B are *similar* over $\mathbb{F}$. If A and B are complex matrices and if P is unitary, i.e., $P^*P = PP^* = I$, we say that A and B are *unitarily similar*.

Similar matrices have the same determinant, for if $B = P^{-1}AP$, then

$$|B| = |P^{-1}AP| = |P^{-1}|\,|A|\,|P| = |P|^{-1}\,|A|\,|P| = |A|.$$

We say a matrix is *diagonalizable* if it is similar to a diagonal matrix and *unitarily diagonalizable* if it is unitarily similar to a diagonalizable matrix.

Trace. Let $A = (a_{ij})$ be an $n \times n$ matrix. The *trace* of A is defined as the sum of the entries on the main diagonal of A; that is,

$$\operatorname{tr} A = a_{11} + a_{22} + \cdots + a_{nn}.$$

Eigenvalues and Eigenvectors of a Matrix. Let A be an $n \times n$ matrix over a field $\mathbb{F}$. A scalar $\lambda \in \mathbb{F}$ is said to be an *eigenvalue* of A if

$$Ax = \lambda x$$

for some nonzero column vector $x \in \mathbb{F}^n$. Such a vector x is referred to as an *eigenvector corresponding* (or *belonging*) *to the eigenvalue* λ.

Let A be an $n \times n$ complex matrix. The fundamental theorem of algebra ensures that A has n complex eigenvalues, including the repeated ones. To see this, observe that $Ax = \lambda x$ is equivalent to $(\lambda I - A)x = 0$, which has a nonzero solution x if and only if $\lambda I - A$ is singular. This is equivalent to λ being a scalar such that $|\lambda I - A| = 0$. Thus, to find the eigenvalues of A, one needs to find the roots of the *characteristic polynomial* of A

$$p_A(\lambda) = |\lambda I - A|.$$

Since the coefficients of $p_A(x)$ are complex numbers, there exist complex numbers $\lambda_1, \lambda_2, \dots, \lambda_n$ (not necessarily distinct) such that

$$p_A(\lambda) = |\lambda I - A| = (\lambda - \lambda_1)(\lambda - \lambda_2) \cdots (\lambda - \lambda_n),$$

so these scalars are the eigenvalues of A.

Expanding the determinant, we see that the constant term of $p_A(\lambda)$ is $(-1)^n|A|$ (this is also seen by putting $\lambda = 0$), and the coefficient of λ is $-\operatorname{tr} A$. Multiplying out the right-hand side and comparing coefficients,

$$|A| = \lambda_1 \lambda_2 \cdots \lambda_n$$

and

$$\operatorname{tr} A = a_{11} + a_{22} + \cdots + a_{nn} = \lambda_1 + \lambda_2 + \cdots + \lambda_n.$$

The eigenvectors x corresponding to the eigenvalue λ are the solutions to the linear equation system $(\lambda I - A)x = 0$; that is, the null space of $\lambda I - A$. We call this space the *eigenspace* of A corresponding to λ.

Similar matrices have the same characteristic polynomial, thus the same eigenvalues and trace but not necessarily the same corresponding eigenvectors. The eigenvalues of an upper- (or lower-) triangular matrix are the elements of the matrix on the main diagonal.

Triangularization and Jordan Canonical Form.

- Let A be an $n \times n$ complex matrix. Then there exists an $n \times n$ invertible complex matrix P such that $P^{-1}AP$ is upper-triangular with the eigenvalues of A on the main diagonal. Simply put: Every square matrix is similar to an upper-triangular matrix over the complex field. Let $\lambda_1, \dots, \lambda_n$ be the eigenvalues of A. We may write

$$P^{-1}AP = \begin{pmatrix} \lambda_1 & & * \\ & \ddots & \\ 0 & & \lambda_n \end{pmatrix}.$$

- Let A be an $n \times n$ complex matrix. Then there exists an $n \times n$ unitary matrix U such that U^*AU is upper-triangular with the eigenvalues of A on the main diagonal. Simply put: Every square matrix is unitarily similar to an upper-triangular matrix over the complex field.

- *Jordan (canonical) form* of a matrix. Let A be an $n \times n$ complex matrix. There exists an $n \times n$ invertible complex matrix P such that

$$P^{-1}AP = \begin{pmatrix} J_1 & 0 & 0 & 0 \\ 0 & J_2 & 0 & 0 \\ \vdots & \vdots & \ddots & \vdots \\ 0 & 0 & 0 & J_s \end{pmatrix},$$

where each J_t, $t = 1, 2, \ldots, s$, called a *Jordan block*, takes the form

$$\begin{pmatrix} \lambda & 1 & 0 & 0 \\ 0 & \lambda & 1 & 0 \\ \vdots & \vdots & \ddots & \vdots \\ 0 & 0 & 0 & \lambda \end{pmatrix}$$

in an appropriate size; λ is an eigenvalue of A. In short: Every square matrix is similar to a matrix in Jordan form over the complex field.

The Jordan form of a matrix carries a great deal of algebraic information about the matrix, and it is useful for solving problems both in theory and computation. For instance, if $\left(\begin{smallmatrix} 2 & 1 \\ 0 & 2 \end{smallmatrix}\right)$ is a Jordan block of a matrix, then this matrix cannot be diagonalizable; that is, it cannot be similar to a diagonal matrix. The determination of the Jordan form of a matrix needs the theory of λ-matrices or generalized eigenvectors. One may find those in many advanced linear algebra books.

Singular Values. Let A be a matrix but not necessarily square. Let λ be an eigenvalue of A^*A and x be a corresponding eigenvector. Then $(A^*A)x = \lambda x$ implies $x^*(A^*A)x = (Ax)^*(Ax) = \lambda x^*x \geq 0$. Hence, $\lambda \geq 0$.

The square roots of the eigenvalues of A^*A are called *singular values* of A. The number of positive singular values of A equals the rank of A.

Let A be an $m \times n$ matrix with rank r, $r \geq 1$, and let $\sigma_1, \sigma_2, \ldots, \sigma_r$ be the positive singular values of A. Then there exist an $m \times m$ unitary (or orthogonal over $\mathbb{R}$) matrix P and an $n \times n$ unitary matrix Q such that

$$A = PDQ,$$

where D is an $m \times n$ matrix with (i,i)-entry σ_i, $i = 1, 2, \ldots, r$, and all other entries 0. This is the well-known *singular value decomposition* theorem.

Linear Transformation. Let V and W be vector spaces over a field $\mathbb{F}$. A mapping $\mathcal{A}$ from V to W is said to be a *linear transformation* if

$$\mathcal{A}(u+v) = \mathcal{A}(u) + \mathcal{A}(v), \qquad u,\ v \in V$$

and

$$\mathcal{A}(ku) = k\mathcal{A}(u), \qquad k \in \mathbb{F},\ u \in V.$$

It follows at once that $\mathcal{A}(0) = 0$. We could have written $\mathcal{A}(0_v) = 0_w$, where 0_v and 0_w stand for the zero vectors of V and W, respectively. However, from the context one can easily tell which is which. For simplicity, we use 0 for both. Sometimes we write $\mathcal{A}(u)$ as $\mathcal{A}u$ for convenience.

The zero transformation from V to W is defined by $0(v) = 0$, $v \in V$.

The linear transformations from V to V are also called *linear operators.*

The Vector Space of Linear Transformations. Let $L(V, W)$ denote the set of all linear transformations from a vector space V to a vector space W. We define addition and scalar multiplication on $L(V, W)$ as follows:

$$(\mathcal{A} + \mathcal{B})(u) = \mathcal{A}(u) + \mathcal{B}(u), \quad (k\mathcal{A})(u) = k(\mathcal{A}(u)).$$

Then $L(V, W)$ is a vector space with respect to the addition and scalar multiplication. The zero vector in $L(V, W)$ is the zero transformation, and for every $\mathcal{A} \in L(V, W)$, $-\mathcal{A}$ is the linear transformation

$$(-\mathcal{A})(u) = -(\mathcal{A}(u)).$$

When $V = W$, $\mathcal{I}(u) = u$, $u \in V$, defines the *identity* transformation on V, and $\mathcal{T}(u) = ku$, $u \in V$, defines a *scalar* transformation for a fixed scalar k. The *product* of linear transformations (operators) $\mathcal{A}$, $\mathcal{B}$ on V can be defined by the composite mapping

$$(\mathcal{A}\mathcal{B})(u) = \mathcal{A}(\mathcal{B}(u)), \qquad u \in V.$$

The product $\mathcal{A}\mathcal{B}$ is once again a linear transformation on V.

Kernel and Image. Let $\mathcal{A}$ be a linear transformation from a vector space V to a vector space W. The *kernel* or *null space* of $\mathcal{A}$ is defined to be

$$\operatorname{Ker} \mathcal{A} = \{\, u \in V \mid \mathcal{A}(u) = 0 \,\}$$

and *image* or *range* of $\mathcal{A}$ is the set

$$\operatorname{Im}\mathcal{A} = \{\, \mathcal{A}(u) \mid u \in V \,\}.$$

The kernel is a subspace of V and the image is a subspace of W. If V is finite dimensional, then both $\operatorname{Ker}\mathcal{A}$ and $\operatorname{Im}\mathcal{A}$ have to be of finite dimension. If $\{u_1, u_2, \ldots, u_s\}$ is a basis for $\operatorname{Ker}\mathcal{A}$ and is extended to a basis for V, $\{u_1, u_2, \ldots, u_s, u_{s+1}, \ldots, u_n\}$, then $\{\mathcal{A}(u_{s+1}), \ldots, \mathcal{A}(u_n)\}$ is a basis for $\operatorname{Im}(\mathcal{A})$. We arrive at the dimension theorem:

$$\dim V = \dim(\operatorname{Ker}\mathcal{A}) + \dim(\operatorname{Im}\mathcal{A}).$$

Given an $m \times n$ matrix A over a field $\mathbb{F}$, we may define a linear transformation from $\mathbb{F}^n$ to $\mathbb{F}^m$ by

$$\mathcal{A}(x) = Ax, \qquad x \in \mathbb{F}^n.$$

The kernel and image of this $\mathcal{A}$ are the null space and column space of A, respectively. As is known from Chapter 1, $\dim(\operatorname{Im}\mathcal{A}) = r(A)$.

Matrix Representation of a Linear Transformation. Let V be a vector space of dimension m with an ordered basis $\alpha = \{\alpha_1, \alpha_2, \ldots, \alpha_m\}$ and W be a vector space of dimension n with an ordered basis $\beta = \{\beta_1, \beta_2, \ldots, \beta_n\}$. If $u \in V$ and $u = x_1\alpha_1 + \cdots + x_m\alpha_m$ for (unique) scalars x_i, letting $x = (x_1, \ldots, x_m)^t$, we will denote this representation of u as αx. Similarly, if $w \in W$ and $w = y_1\beta_1 + \cdots + y_n\beta_n$, we will abbreviate as $w = \beta y$.

Let $\mathcal{A}$ be a linear transformation from V to W. Then $\mathcal{A}$ is determined by its action on the ordered basis α relative to β. To be precise, let

$$\mathcal{A}(\alpha_i) = a_{1i}\beta_1 + a_{2i}\beta_2 + \cdots + a_{ni}\beta_n, \quad i = 1, 2, \ldots, m.$$

For the sake of convenience, we use the following notation:

$$\mathcal{A}(\alpha_i) = (\beta_1, \beta_2, \ldots, \beta_n)a_i = \beta a_i, \quad \text{where } a_i = (a_{1i},\, a_{2i},\, \ldots,\, a_{ni})^t$$

and

$$\mathcal{A}(\alpha) = (\mathcal{A}(\alpha_1), \mathcal{A}(\alpha_2), \ldots, \mathcal{A}(\alpha_m)) = (\beta_1, \beta_2, \ldots, \beta_n)A = \beta A,$$

where

$$A = (a_1, a_2, \ldots, a_n) = (a_{ij}).$$

If

$$u = x_1\alpha_1 + x_2\alpha_2 + \cdots + x_m\alpha_m = (\alpha_1, \alpha_2, \ldots, \alpha_m)x = \alpha x,$$

where $x = (x_1, x_2, \ldots, x_m)^t$ is the coordinate of u relative to basis α, then

$$\mathcal{A}(u) = \mathcal{A}(\alpha x) = (\mathcal{A}(\alpha))x = (\beta A)x = \beta(Ax).$$

This says Ax is the coordinate vector of $\mathcal{A}(u) \in W$ relative to the basis β. Thus the linear transformation $\mathcal{A}$ is determined by the matrix A. Such a matrix A associated to $\mathcal{A}$ is called the *matrix representation* of the linear transformation $\mathcal{A}$ relative to the (ordered) bases α of V and β of W.

If $V = W$ and $\alpha = \beta$, then $\mathcal{A}(\alpha) = \alpha A$; we simply say that A is the matrix of $\mathcal{A}$ under, or relative to, the basis α. If $V = \mathbb{F}^m$ and $W = \mathbb{F}^n$, with the standard bases $\alpha = \{e_1, \ldots, e_m\}$ and $\beta = \{\epsilon_1, \ldots, \epsilon_n\}$, we have

$$\mathcal{A}(u) = Ax.$$

Matrices of a Linear Operator Are Similar. Consider V, a vector space of dimension n. Let α and β be two bases for V. Then there exists an n-square invertible matrix P such that $\beta = \alpha P$. Let A_1 be the matrix of $\mathcal{A}$ under the basis α; that is, $\mathcal{A}(\alpha) = \alpha A_1$. Let A_2 be the matrix under β. We claim that A_1 and A_2 are similar. This is because

$$\mathcal{A}(\beta) = \mathcal{A}(\alpha P) = (\mathcal{A}(\alpha))P = (\alpha A_1)P = \beta(P^{-1}A_1P).$$

It follows that $A_2 = P^{-1}A_1P$. Therefore, the matrices of a linear operator under different bases are similar.

Eigenvalues of a Linear Operator. Let $\mathcal{A}$ be a linear transformation on a vector space V over $\mathbb{F}$. A scalar $\lambda \in \mathbb{F}$ is an *eigenvalue* of $\mathcal{A}$ if $\mathcal{A}(u) = \lambda u$ for some nonzero vector u. Such a vector u is called an *eigenvector* of $\mathcal{A}$ corresponding to the eigenvalue λ.

Let A be the matrix of $\mathcal{A}$ under a basis α of V and x be the coordinate of the vector u under α; that is, $u = \alpha x$. Then

$$\alpha(\lambda x) = \lambda(\alpha x) = \lambda u = \mathcal{A}(u) = \mathcal{A}(\alpha x) = (\mathcal{A}(\alpha))x = (\alpha A)x = \alpha(Ax).$$

Thus $\mathcal{A}(u) = \lambda u$ is equivalent to $Ax = \lambda x$. So the eigenvalues of the linear transformation $\mathcal{A}$ are just the eigenvalues of its matrix A under α. Note that similar matrices have the same eigenvalues. The eigenvalues of $\mathcal{A}$ through its matrices are independent of the choices of the bases.

Invariant Subspaces. Let $\mathcal{A}$ be a linear operator on a vector space V. If W is a subspace of V such that $\mathcal{A}(w) \in W$ for all $w \in W$; that is, $\mathcal{A}(W) \subseteq W$, then we say that W is *invariant* under $\mathcal{A}$. Both $\operatorname{Ker} \mathcal{A}$ and $\operatorname{Im} \mathcal{A}$ are invariant subspaces under any linear operator $\mathcal{A}$.

Chapter 3 Problems

3.1 Let A and B be n-square matrices. Answer true or false:

(a) If $A^2 = 0$, then $A = 0$.

(b) If $A^2 = 0$ and λ is an eigenvalue of A, then $\lambda = 0$.

(c) If $A^2 = 0$, then the rank of A is at most 2.

(d) If $A^2 = A$, then $A = 0$ or I.

(e) If $A^*A = 0$, then $A = 0$.

(f) If $AB = 0$, then $A = 0$ or $B = 0$.

(g) If $|AB| = 0$, then $|A| = 0$ or $|B| = 0$.

(h) $AB = BA$.

(i) $|AB| = |BA|$, where A is $m \times n$ and B is $n \times m$.

(j) $|A + B| = |A| + |B|$.

(k) $(A + I)^2 = A^2 + 2A + I$.

(l) $|kA| = k|A|$ for any scalar k.

3.2 Let A and B be $n \times n$ matrices. Show that

$$(A + B)^2 = A^2 + 2AB + B^2$$

if and only if A and B commute; that is, $AB = BA$.

3.3 Let A and B be $n \times n$ matrices. Show that

$$AB = A \pm B \quad \Rightarrow \quad AB = BA.$$

3.4 Find the values of a and b such that the following matrices are similar:

$$A = \begin{pmatrix} -2 & 0 & 0 \\ 2 & a & 2 \\ 3 & 1 & 1 \end{pmatrix}, \quad B = \begin{pmatrix} -1 & 0 & 0 \\ 0 & 2 & 0 \\ 0 & 0 & b \end{pmatrix}.$$

3.5 What are the matrices that are similar to themselves only?

3.6 A matrix X is said to be *equivalent* to matrix A if $PXQ = A$ for some invertible matrices P and Q; *congruent* to A if $P^t XP = A$ for some invertible P; and *similar* to A if $P^{-1}XP = A$ for some invertible P.

Let A be the diagonal matrix $\operatorname{diag}(1, 2, -1)$. Determine if the matrices

$$B = \begin{pmatrix} 1 & -1 & 0 \\ -1 & 2 & 0 \\ 0 & 0 & 3 \end{pmatrix}, \quad C = \begin{pmatrix} -2 & 0 & 0 \\ 0 & 1 & 0 \\ 0 & 0 & 1 \end{pmatrix}, \quad D = \begin{pmatrix} 0 & 1 & 0 \\ 1 & 0 & 0 \\ 0 & 0 & 2 \end{pmatrix}$$

are

(a) equivalent to A;

(b) congruent to A; or

(c) similar to A.

3.7 Which of the following matrices are similar to $A = \operatorname{diag}(1, 4, 6)$?

$$B = \begin{pmatrix} 1 & 2 & 3 \\ 0 & 4 & 5 \\ 0 & 0 & 6 \end{pmatrix}, \quad C = \begin{pmatrix} 4 & 0 & 0 \\ 7 & 1 & 0 \\ 8 & 9 & 6 \end{pmatrix}, \quad D = \begin{pmatrix} 1 & 2 & 0 \\ 3 & 4 & 5 \\ 0 & 7 & 6 \end{pmatrix},$$

$$E = \begin{pmatrix} 4 & 7 & 0 \\ 0 & 1 & 0 \\ 8 & 9 & 6 \end{pmatrix}, \quad F = \begin{pmatrix} 1 & 2 & 0 \\ 3 & 4 & 0 \\ 0 & 5 & 6 \end{pmatrix}, \quad G = \begin{pmatrix} 1 & 0 & 0 \\ 2 & 5 & 1 \\ 3 & 1 & 5 \end{pmatrix}.$$

3.8 Let $a, b, c \in \mathbb{R}$. Find the condition on a, b, and c such that the matrix

$$\begin{pmatrix} 2 & 0 & 0 \\ a & 2 & 0 \\ b & c & -1 \end{pmatrix}$$

is similar to a diagonal matrix.

3.9 For any scalars a, b, and c, show that

$$A = \begin{pmatrix} b & c & a \\ c & a & b \\ a & b & c \end{pmatrix}, \quad B = \begin{pmatrix} c & a & b \\ a & b & c \\ b & c & a \end{pmatrix}, \quad C = \begin{pmatrix} a & b & c \\ b & c & a \\ c & a & b \end{pmatrix}$$

are similar. Moreover, if $BC = CB$, then A has two zero eigenvalues.

3.10 Let E_{ij} be the n-square matrix with the (i, j)-entry 1 and 0 elsewhere, $i, j = 1, 2, \ldots, n$. For $A \in M_n(\mathbb{C})$, find AE_{ij}, $E_{ij}A$, and $E_{ij}AE_{st}$.

3.11 Compute A^2 and A^6, where

$$A = \begin{pmatrix} -1 & 1 & 1 & -1 \\ 1 & -1 & -1 & 1 \\ 1 & -1 & -1 & 1 \\ -1 & 1 & 1 & -1 \end{pmatrix}.$$

3.12 Find A^{100}, where

$$A = \begin{pmatrix} 1 & 2 \\ 3 & 4 \end{pmatrix}.$$

3.13 For positive integer $k \geq 2$, compute

$$\begin{pmatrix} 2 & 1 \\ 2 & 3 \end{pmatrix}^k, \quad \begin{pmatrix} \lambda & 1 \\ 0 & \lambda \end{pmatrix}^k, \quad \begin{pmatrix} 0 & 1 & 0 \\ 0 & 0 & 1 \\ 0 & 0 & 0 \end{pmatrix}^k, \quad \begin{pmatrix} 0 & 1 & 0 \\ 0 & 0 & 1 \\ 1 & 0 & 0 \end{pmatrix}^k.$$

3.14 Let $A = \left(\begin{smallmatrix} 1 & 1 \\ 0 & 1 \end{smallmatrix}\right)$. Show that A^k is similar to A for every positive integer k. This is true more generally for any matrix with all eigenvalues 1.

3.15 Let $u = (1, 2, 3)$ and $v = (1, \frac{1}{2}, \frac{1}{3})$. Let $A = u^t v$. Find A^n, $n \geq 1$.

3.16 Let A be an $n \times n$ complex matrix. Show that

(a) (Schur Decomposition) There is a unitary matrix U such that

$$U^*AU = \begin{pmatrix} \lambda_1 & & & * \\ & \lambda_2 & & \\ & & \ddots & \\ 0 & & & \lambda_n \end{pmatrix}$$

is an upper-triangular matrix, where λ_i's are the eigenvalues of A, and $*$ represents unspecified entries.

(b) If A and $B \in M_n(\mathbb{C})$ are similar, then for any polynomial $f(x)$ in x, $f(A)$ and $f(B)$ are similar.

(c) If λ is an eigenvalue of A, then $f(\lambda)$ is an eigenvalue of $f(A)$. In particular, λ^k is an eigenvalue of A^k.

(d) If $AP = QA$ for diagonal P and Q, then $Af(P) = f(Q)A$.

3.17 Show that an n-square matrix is similar to a diagonal matrix if and only if the matrix has n linearly independent eigenvectors. Does the matrix have to have n distinct eigenvalues?

3.18 Let A be a square matrix such that $|A| = 0$. Show that there exists a positive number δ such that $|A + \epsilon I| \neq 0$, for any $\epsilon \in (0, \delta)$.

3.19 Show that for any 2×2 matrix A and 3×3 matrix B,

$$A^2 - (\operatorname{tr} A)A + |A|I = 0$$

and

$$|\lambda I - B| = \lambda^3 - \lambda^2 \operatorname{tr} B + \lambda \operatorname{tr}(\operatorname{adj}(B)) - |B|.$$

3.20 Let $A, B \in M_n(\mathbb{C})$ and let

$$p_B(\lambda) = |\lambda I - B|$$

be the characteristic polynomial of B. Show that the matrix $p_B(A)$ is invertible if and only if A and B have no common eigenvalues.

3.21 Let $B \in M_n(\mathbb{C})$, u and v be $1 \times n$ and $n \times 1$ vectors, respectively. Let

$$A = \begin{pmatrix} B & -Bv \\ -uB & uBv \end{pmatrix}.$$

(a) Show that $|A| = 0$.

(b) If $|B| = 0$, then λ^2 divides $|\lambda I - A|$.

(c) Discuss the converse of (b).

3.22 Let A and B be real matrices such that $A + iB$ is nonsingular. Show that there exists a real number t such that $A + tB$ is nonsingular.

3.23 Let A and B be n-square matrices. Show that the characteristic polynomial of the following matrix M is an even function; that is,

$$\text{if} \quad |\lambda I - M| = 0, \quad \text{then} \quad |-\lambda I - M| = 0,$$

where

$$M = \begin{pmatrix} 0 & A \\ B & 0 \end{pmatrix}.$$

3.24 Let A and B be n-square matrices. Answer true or false:

(a) If $A^k = 0$ for all positive integers $k \geq 2$, then $A = 0$.

(b) If $A^k = 0$ for some positive integer k, then $\operatorname{tr} A = 0$.

(c) If $A^k = 0$ for some positive integer k, then $|A| = 0$.

(d) If $A^k = 0$ for some positive integer k, then $r(A) = 0$.

(e) If $\operatorname{tr} A = 0$, then $|A| = 0$.

(f) If the rank of A is r, then A has r nonzero eigenvalues.

(g) If A has r nonzero eigenvalues, then $r(A) \geq r$.

(h) If A and B are similar, then they have the same eigenvalues.

(i) If A and B are similar, then they have the same singular values.

(j) If A and B have the same eigenvalues, then they are similar.

(k) If A and B have the same characteristic polynomial, then they have the same eigenvalues; and vice versa.

(l) If A and B have the same characteristic polynomial, then they are similar.

(m) If all eigenvalues of A are zero, then $A = 0$.

(n) If all singular values of A are zero, then $A = 0$.

(o) If $\operatorname{tr} A^k = \operatorname{tr} B^k$ for all positive integers k, then $A = B$.

(p) If the eigenvalues of A are $\lambda_1, \lambda_2, \ldots, \lambda_n$, then A is similar to the diagonal matrix $\operatorname{diag}(\lambda_1, \lambda_2, \ldots, \lambda_n)$.

(q) $\operatorname{diag}(1, 2, \ldots, n)$ is similar to $\operatorname{diag}(n, \ldots, 2, 1)$.

(r) If A has a repeated eigenvalue, then A is not diagonalizable.

(s) If $a + bi$ is an eigenvalue of a real square matrix A, then $a - bi$ is also an eigenvalue of the matrix A.

(t) If A is a real square matrix, then all eigenvalues of A are real.

3.25 Let $A \in M_n(\mathbb{C})$. Prove assertions (a) and (b):

(a) If the eigenvalues of A are distinct from each other, then A is diagonalizable; that is, there is an invertible matrix P such that $P^{-1}AP$ is a diagonal matrix.

(b) If matrix A commutes with a matrix with all distinct eigenvalues, then A is diagonalizable.

3.26 Let A be an $n \times n$ nonsingular matrix having distinct eigenvalues. If B is a matrix satisfying $AB = BA^{-1}$, show that B^2 is diagonalizable.

3.27 Show that if all the eigenvalues of $A \in M_n(\mathbb{C})$ are real and if

$$\operatorname{tr} A^2 = \operatorname{tr} A^3 = \operatorname{tr} A^4 = c$$

for some constant c, then for every positive integer k,

$$\operatorname{tr} A^k = c,$$

and c must be an integer. The same conclusion can be drawn if $A^m = A^{m+1}$ for some positive integer m.

3.28 Let $A \in M_n(\mathbb{C})$. Show that

$$A^n = 0 \text{ if } \operatorname{tr} A^k = 0, \ k = 1, 2, \ldots, n.$$

3.29 Let $A, B \in M_n(\mathbb{C})$. If $AB = 0$, show that for any positive integer k,

$$\operatorname{tr}(A + B)^k = \operatorname{tr} A^k + \operatorname{tr} B^k.$$

3.30 Let $A = \left(\begin{smallmatrix} 0 & 1 \\ 1 & 1 \end{smallmatrix}\right)$. Show that for any positive integer $k \geq 2$

$$\operatorname{tr} A^k = \operatorname{tr} A^{k-1} + \operatorname{tr} A^{k-2}.$$

3.31 Find the eigenvalues and corresponding eigenvectors of the matrix

$$A = \begin{pmatrix} 1 & 2 & 2 \\ 2 & 1 & 2 \\ 2 & 2 & 1 \end{pmatrix}.$$

And then find an invertible matrix P such that $P^{-1}AP$ is diagonal.

3.32 Show that the following matrix is not similar to a diagonal matrix:

$$A = \begin{pmatrix} 2 & 3 & 2 \\ 1 & 4 & 2 \\ 1 & -3 & 1 \end{pmatrix}.$$

3.33 If matrix

$$A = \begin{pmatrix} 0 & 0 & 1 \\ x & 1 & y \\ 1 & 0 & 0 \end{pmatrix}$$

has three linearly independent eigenvectors, show that $x + y = 0$.

3.34 If matrices

$$A = \begin{pmatrix} 1 & a & 1 \\ a & 1 & b \\ 1 & b & 1 \end{pmatrix} \quad \text{and} \quad B = \begin{pmatrix} 0 & 0 & 0 \\ 0 & 1 & 0 \\ 0 & 0 & 2 \end{pmatrix}$$

are similar, what are the values of a and b? Find a real orthogonal matrix T, namely, $T^tT = TT^t = I$, such that $T^{-1}AT = B$.

3.35 Let λ be an eigenvalue of an n-square matrix $A = (a_{ij})$. Show that there exists a positive integer k such that

$$|\lambda - a_{kk}| \leq \sum_{j=1,\, j\neq k}^{n} |a_{kj}|.$$

3.36 If the eigenvalues of $A = (a_{ij}) \in M_n(\mathbb{C})$ are $\lambda_1, \lambda_2, \ldots, \lambda_n$, show that

$$\sum_{i=1}^{n} |\lambda_i|^2 \leq \sum_{i,\, j=1}^{n} |a_{ij}|^2$$

and equality holds if and only if A is unitarily diagonalizable.

3.37 Let A be an n-square real matrix with eigenvalues $\lambda_1, \lambda_2, \ldots, \lambda_n$ (which are not necessarily real). Denote $\lambda_k = x_k + iy_k$. Show that

(a) $y_1 + y_2 + \cdots + y_n = 0$.

(b) $x_1y_1 + x_2y_2 + \cdots + x_ny_n = 0$.

(c) $\operatorname{tr} A^2 = (x_1^2 + x_2^2 + \cdots + x_n^2) - (y_1^2 + y_2^2 + \cdots + y_n^2)$.

3.38 Let λ_1 and λ_2 be two different eigenvalues of a matrix A and let u_1 and u_2 be eigenvectors of A corresponding to λ_1 and λ_2, respectively. Show that $u_1 + u_2$ is not an eigenvector of A.

3.39 Find a 3×3 real matrix A such that

$$Au_1 = u_1, \quad Au_2 = 2u_2, \quad Au_3 = 3u_3,$$

where

$$u_1 = \begin{pmatrix} 1 \\ 2 \\ 2 \end{pmatrix}, \quad u_2 = \begin{pmatrix} 2 \\ -2 \\ 1 \end{pmatrix}, \quad u_3 = \begin{pmatrix} -2 \\ -1 \\ 2 \end{pmatrix}.$$

3.40 If $A \in M_2(\mathbb{R})$ satisfies $A^2 + I = 0$, show that A is similar to $\left(\begin{smallmatrix} 0 & -1 \\ 1 & 0 \end{smallmatrix}\right)$.

3.41 Let $A = \left(\begin{smallmatrix} a & b \\ c & d \end{smallmatrix}\right)$ be a 2×2 real matrix. If $\left(\begin{smallmatrix} x_0 \\ 1 \end{smallmatrix}\right)$ is an eigenvector of A for some eigenvalue, find the value of x_0 in terms of a, b, c, and d.

3.42 Let $A = \left(\begin{smallmatrix} a & b \\ c & d \end{smallmatrix}\right) \in M_2(\mathbb{C})$ and $|A| = 1$.

(a) Find A^{-1}.

(b) Write A as a product of matrices of the forms

$$\begin{pmatrix} 1 & x \\ 0 & 1 \end{pmatrix} \quad \text{and} \quad \begin{pmatrix} 1 & 0 \\ x & 1 \end{pmatrix}.$$

(c) If $|a+d| > 2$, then A is similar to $\left(\begin{smallmatrix} \lambda & 0 \\ 0 & \lambda^{-1} \end{smallmatrix}\right)$, where $\lambda \neq 0,\ 1,\ -1$.

(d) If $|a+d| < 2$, then A is similar to $\left(\begin{smallmatrix} \lambda & 0 \\ 0 & \lambda^{-1} \end{smallmatrix}\right)$, where $\lambda \notin \mathbb{R} \cup i\mathbb{R}$.

(e) If $|a+d| = 2$ and A has real eigenvalues, what are the possible real matrices to which A is similar?

(f) If $|a+d| \neq 2$, then A is similar to $\left(\begin{smallmatrix} \frac{a+d}{2} & x \\ x & \frac{a+d}{2} \end{smallmatrix}\right)$ for some $x \in \mathbb{C}$.

(g) Does (f) remain true if $|a+d| = 2$?

3.43 Show that matrices A and B are similar but not unitarily similar:

$$A = \begin{pmatrix} 1 & 1 \\ 0 & 0 \end{pmatrix}, \quad B = \begin{pmatrix} 1 & 0 \\ 0 & 0 \end{pmatrix}.$$

3.44 Show that if two real square matrices are similar over $\mathbb{C}$, then they must be similar over $\mathbb{R}$. What if "real" is changed to "rational"? Find a rational matrix M such that

$$M^{-1}\begin{pmatrix} 1 & 2 \\ 2 & -1 \end{pmatrix} M = \begin{pmatrix} 2 & 1 \\ 1 & -2 \end{pmatrix}.$$

3.45 Find the eigenvalues and corresponding eigenvectors of the matrix

$$\begin{pmatrix} 2 & 1 & 0 \\ 1 & 3 & 1 \\ 0 & 1 & 2 \end{pmatrix}.$$

Show that the eigenvectors of distinct eigenvalues are orthogonal.

3.46 Find a singular value decomposition (SVD) for the 3×2 matrix

$$A = \begin{pmatrix} 1 & 0 \\ 0 & 1 \\ 1 & 0 \end{pmatrix}.$$

3.47 Show that $T^{-1}AT$ is always diagonal for any numbers x and y, where

$$A = \begin{pmatrix} x & y & 0 \\ y & x & y \\ 0 & y & x \end{pmatrix}, \quad T = \begin{pmatrix} \frac{1}{2} & \frac{\sqrt{2}}{2} & \frac{1}{2} \\ \frac{\sqrt{2}}{2} & 0 & -\frac{\sqrt{2}}{2} \\ \frac{1}{2} & -\frac{\sqrt{2}}{2} & \frac{1}{2} \end{pmatrix}.$$

3.48 For any $n \times n$ complex matrix A, show that A and A^t are similar. Are A and A^* necessarily similar? Can A be similar to $A + I$?

3.49 If A is a singular square matrix, what are the eigenvalues of $\operatorname{adj}(A)$?

3.50 Let A be an $n \times n$ complex matrix. Show that $\lambda \geq 0$ is an eigenvalue of $A\bar{A}$ if and only if $A\bar{x} = \sqrt{\lambda}\, x$ for some nonzero $x \in \mathbb{C}^n$.

3.51 Let A, B be $m \times n$, $n \times m$ matrices, respectively, $m \geq n$. Show that

$$|\lambda I_m - AB| = \lambda^{m-n} |\lambda I_n - BA|.$$

Conclude that AB and BA have the same nonzero eigenvalues, counting multiplicities. Do they have the same singular values?

3.52 Let $a_1, a_2, \ldots, a_n \in \mathbb{R}$ be such that $a_1 + a_2 + \cdots + a_n = 0$ and denote

$$A = \begin{pmatrix} a_1^2+1 & a_1a_2+1 & \ldots & a_1a_n+1 \\ a_2a_1 & a_2^2+1 & \ldots & a_2a_n+1 \\ \ldots & \ldots & \ldots & \ldots \\ a_na_1+1 & a_na_2+1 & \ldots & a_n^2+1 \end{pmatrix}.$$

Show that $A = BB^t$ for some matrix B. Find the eigenvalues of A.

3.53 Let $u, v \in \mathbb{R}^n$ be nonzero column vectors orthogonal to each other; that is, $v^tu = 0$. Find all eigenvalues of $A = uv^t$ and corresponding eigenvectors. Find also A^2. Is A similar to a diagonal matrix?

3.54 Let A and B be square matrices of the same size. Show that matrices M and P are similar to matrices N and Q, respectively, where

$$M = \begin{pmatrix} A & B \\ B & A \end{pmatrix}, \quad N = \begin{pmatrix} A+B & 0 \\ 0 & A-B \end{pmatrix},$$

$$P = \begin{pmatrix} A & -B \\ B & A \end{pmatrix}, \quad Q = \begin{pmatrix} A+iB & 0 \\ 0 & A-iB \end{pmatrix}.$$

3.55 Let $A,\ B \in M_n(\mathbb{C})$.

(a) Show that $\operatorname{tr}(AB) = \operatorname{tr}(BA)$.
(b) Show that $\operatorname{tr}(AB)^k = \operatorname{tr}(BA)^k$.
(c) Is it true that $\operatorname{tr}(AB)^k = \operatorname{tr}(A^kB^k)$?
(d) Why is A singular if $AB - BA = A$?
(e) Show that $\operatorname{tr}(ABC) = \operatorname{tr}(BCA)$ for every $C \in M_n(\mathbb{C})$.
(f) Is it true that $\operatorname{tr}(ABC) = \operatorname{tr}(ACB)$?
(g) Show that $\operatorname{tr}[(AB - BA)(AB + BA)] = 0$.
(h) Show that AB and BA are similar if A or B is nonsingular.
(i) Are AB and BA similar in general?

3.56 Let J_n denote the n-square matrix all of whose entries are 1. Find the eigenvalues and corresponding eigenvectors of J_n. Let

$$K = \begin{pmatrix} 0 & J_n \\ J_n & 0 \end{pmatrix}.$$

Find the eigenvalues and corresponding eigenvectors of K.

3.57 Let A be the $n \times n$ matrix all of whose main diagonal entries are 0 and elsewhere 1, i.e., $a_{ii} = 0$, $1 \le i \le n$ and $a_{ij} = 1$, $i \ne j$. Find A^{-1}.

3.58 Let $A \in M_n(\mathbb{C})$ be a matrix with real eigenvalues, and let s be the number of nonzero eigenvalues of A. Show that

(a) $(\operatorname{tr} A)^2 \le s \operatorname{tr} A^2$. When does equality hold?

(b) $(\operatorname{tr} A)^2 \le r(A) \operatorname{tr} A^2$ when A is Hermitian. Moreover, equality holds if and only if $A^2 = cA$ for some scalar c.

(c) If $(\operatorname{tr} A)^2 > (n-1) \operatorname{tr} A^2$, then A is nonsingular.

3.59 Let $A \in M_n(\mathbb{C})$. Show that if $A^3 = A$, then $r(A) = \operatorname{tr} A^2$.

3.60 Let m and j be positive integers with $m \ge j$. Let $S_{m,j}(X, Y)$ denote the sum of all matrix products of the form $A_1 \cdots A_m$, where each A_i is either X or Y, and is Y in exactly j cases. Show that

$$\operatorname{tr}(S_{5,3}(X, Y)) = \frac{5}{2} \operatorname{tr}(X S_{4,3}(X, Y)).$$

3.61 If A and B are 3×2 and 2×3 matrices, respectively, such that

$$AB = \begin{pmatrix} 8 & 2 & -2 \\ 2 & 5 & 4 \\ -2 & 4 & 5 \end{pmatrix},$$

show that

$$BA = \begin{pmatrix} 9 & 0 \\ 0 & 9 \end{pmatrix}.$$

3.62 Let A be a 3×3 real symmetric matrix. It is known that 1, 2, and 3 are the eigenvalues of A and that $\alpha_1 = (-1, -1, 1)^t$ and $\alpha_2 = (1, -2, -1)^t$ are eigenvectors of A belonging to the eigenvalues 1 and 2, respectively. Find an eigenvector of A corresponding to the eigenvalue 3 and then find the matrix A.

3.63 Construct a 3×3 real symmetric matrix A such that the eigenvalues of A are 1, 1, and -1, and $\alpha = (1, 1, 1)^t$ and $\beta = (2, 2, 1)^t$ are eigenvectors corresponding to the eigenvalue 1.

3.64 For $A, B \in M_n(\mathbb{C})$, $AB - BA$ is called the *commutator* of A and B, and it is denoted by $[A, B]$. Show that

(a) $[A, B] = [-A, -B] = -[B, A]$.

(b) $[A, B + C] = [A, B] + [A, C]$.

(c) $[A, B]^* = [B^*, A^*]$.

(d) $[PXP^{-1}, Y] = 0$ if and only if $[X, P^{-1}YP] = 0$.

(e) $\operatorname{tr}[A, B] = 0$.

(f) $I - [A, B]$ is not nilpotent.

(g) $[A, B]$ is never similar to the identity matrix.

(h) If the diagonal entries of A are all equal to zero, then there exist matrices X and Y such that $A = [X, Y]$.

(i) If $[A, B] = 0$, then $[A^p, B^q] = 0$ for all positive integers p, q.

(j) If $[A, B] = A$, then A is singular.

(k) If A and B are both Hermitian or skew-Hermitian, then $[A, B]$ is skew-Hermitian.

(l) If one of A and B is Hermitian and the other one is skew-Hermitian, then $[A, B]$ is Hermitian.

(m) If A is a skew-Hermitian matrix, then $A = [B, C]$ for some Hermitian matrices B and C.

(n) If A and B are Hermitian, then the real part of every eigenvalue of $[A, B]$ is zero.

(o) If $[A, [A, A^*]] = 0$, then A is normal.

(p) $[A, [B, C]] + [B, [C, A]] + [C, [A, B]] = 0$.

(q) If $[A, B]$ commutes with A and B, then $[A, B]$ has no eigenvalues other than 0, and further $[A, B]^k = 0$ for some k.

When does it happen that $[A, B] = [B, A]$?

3.65 Show that $A \in M_n(\mathbb{C})$ is diagonalizable, meaning $P^{-1}AP$ is diagonal for some invertible P, if and only if for every eigenvalue λ of A,

$$r(A - \lambda I) = r[(A - \lambda I)^2].$$

Equivalently, A is diagonalizable if and only if $(\lambda I - A)x = 0$ whenever $(\lambda I - A)^2 x = 0$, where x is a column vector of n components.

3.66 Let $A \in M_n(\mathbb{C})$. Show that the following are equivalent:

(a) $A^2 = BA$ for some nonsingular matrix B.

(b) $r(A^2) = r(A)$.

(c) $\operatorname{Im} A \cap \operatorname{Ker} A = \{0\}$.

(d) There exist nonsingular matrices P and D of orders $n \times n$ and $r(A) \times r(A)$, respectively, such that

$$A = P \begin{pmatrix} D & 0 \\ 0 & 0 \end{pmatrix} P^{-1}.$$

3.67 Let $A \in M_n(\mathbb{C})$. Of the matrices $\bar{A}$, A^t, A^*, $\operatorname{adj}(A)$, $\frac{A+A^*}{2}$, $(A^*A)^{\frac{1}{2}}$, which always has the same eigenvalues or singular values as A?

3.68 Let A be a square complex matrix and denote

$$\begin{aligned} \rho &= \max\{\, |\lambda| \mid \lambda \text{ is an eigenvalue of } A \,\}, \\ \omega &= \max\{\, |x^*Ax| \mid x^*x = 1 \,\}, \\ \sigma &= \max\{\, (x^*A^*Ax)^{1/2} \mid x^*x = 1 \,\}. \end{aligned}$$

Show that

$$\rho \le \omega \le \sigma.$$

3.69 Let A, B, and C be $n \times n$ complex matrices. Show that

$$AB = AC \quad \text{if and only if} \quad A^*AB = A^*AC.$$

3.70 Let A and B be $n \times n$ complex matrices of the same rank. Show that

$$A^2B = A \quad \text{if and only if} \quad B^2A = B.$$

3.71 Let $A = I - (x^*x)^{-1}(xx^*)$, where x is a nonzero n-column vector. Find

(a) $r(A)$. (b) $\operatorname{Im} A$. (c) $\operatorname{Ker} A$.

3.72 If $A \in M_n(\mathbb{Q})$, show that there exists a polynomial $f(x)$ of integer coefficients such that $f(A) = 0$. Find such a polynomial $f(x)$ of the lowest degree for which $f(A) = 0$, where $A = \operatorname{diag}(\frac{1}{2}, \frac{2}{3}, \frac{3}{4})$.

3.73 Let A and B be m-square and n-square matrices, respectively. If A and B have no common eigenvalue over $\mathbb{C}$, show that the matrix equation $AX = XB$ will have only the zero solution $X = 0$.

3.74 Prove

(a) If $\lambda \neq 0$ is an eigenvalue of A, $\frac{1}{\lambda}|A|$ is an eigenvalue of $\operatorname{adj}(A)$.

(b) If v is an eigenvector of A, v is also an eigenvector of $\operatorname{adj}(A)$.

3.75 Let A and B be $n \times n$ matrices such that $AB = BA$. Show that

(a) If A has n distinct eigenvalues, then A, B, and AB are all diagonalizable.

(b) If A and B are diagonalizable, then there exists an invertible matrix T such that $T^{-1}AT$ and $T^{-1}BT$ are both diagonal.

3.76 Which of the following $\mathcal{A}$ are linear transformations on $\mathbb{C}^n$?

(a) $\mathcal{A}(u) = v$, where $v \neq 0$ is a fixed vector in $\mathbb{C}^n$.

(b) $\mathcal{A}(u) = 0$.

(c) $\mathcal{A}(u) = \bar{u}$.

(d) $\mathcal{A}(u) = ku$, where k is a fixed complex number.

(e) $\mathcal{A}(u) = \|u\|\, u$, where $\|u\|$ is the length of vector u.

(f) $\mathcal{A}(u) = u + v$, where $v \neq 0$ is a fixed vector in $\mathbb{C}^n$.

(g) $\mathcal{A}(u) = (u_1, 2u_2, \ldots, nu_n)$, where $u = (u_1, u_2, \ldots, u_n)$.

3.77 Let $\mathcal{A}$ be a linear transformation on a vector space. Show that

$$\operatorname{Ker}\mathcal{A} \subseteq \operatorname{Im}(\mathcal{I} - \mathcal{A})$$

and

$$\operatorname{Im}\mathcal{A} \subseteq \operatorname{Ker}(\mathcal{I} - \mathcal{A}).$$

3.78 Let A, B, C, D be $n \times n$ complex matrices. Define $\mathcal{T}$ on $M_n(\mathbb{C})$ by

$$\mathcal{T}(X) = AXB + CX + XD, \quad X \in M_n(\mathbb{C}).$$

Show that $\mathcal{T}$ is a linear transformation on $M_n(\mathbb{C})$ and that when $C = D = 0$, $\mathcal{T}$ has an inverse if and only if A and B are invertible.

3.79 Let $A \in M_n(\mathbb{C})$ and $A \neq 0$. Define a transformation on $M_n(\mathbb{C})$ by

$$\mathcal{T}(X) = AX - XA, \quad X \in M_n(\mathbb{C}).$$

Show that

(a) $\mathcal{T}$ is linear.

(b) Zero is an eigenvalue of $\mathcal{T}$.

(c) If $A^k = 0$, then $\mathcal{T}^{2k} = 0$.

(d) $\mathcal{T}(XY) = X\mathcal{T}(Y) + \mathcal{T}(X)Y$.

(e) If A is diagonalizable, so is a matrix representation of $\mathcal{T}$.

(f) If A and B commute, so do $\mathcal{T}$ and $\mathcal{L}$, where $\mathcal{L}$ is defined as

$$\mathcal{L}(X) = BX - XB, \quad X \in M_n(\mathbb{C}).$$

Find all A such that $\mathcal{T} = 0$, and discuss the converse of (f).

3.80 Let $\mathcal{A}$ be a linear transformation from a vector space V to a vector space W and $\dim V = n$. If

$$\{\alpha_1, \ldots, \alpha_s, \alpha_{s+1}, \ldots, \alpha_n\}$$

is a basis for V such that $\{\alpha_1, \ldots, \alpha_s\}$ is a basis for $\operatorname{Ker} \mathcal{A}$, show that

(a) $\{\mathcal{A}(\alpha_{s+1}), \ldots, \mathcal{A}(\alpha_n)\}$ is a basis for $\operatorname{Im} \mathcal{A}$.

(b) $\dim(\operatorname{Ker} \mathcal{A}) + \dim(\operatorname{Im} \mathcal{A}) = n$.

(c) $V = \operatorname{Ker} \mathcal{A} \oplus \operatorname{Span}\{\alpha_{s+1}, \ldots, \alpha_n\}$.

Is $\operatorname{Ker} \mathcal{A} + \operatorname{Im} \mathcal{A}$ necessarily a direct sum when $V = W$? If $\{\beta_1, \ldots, \beta_n\}$ is a basis for V, does it necessarily follow that some β_i's fall in $\operatorname{Ker} \mathcal{A}$?

3.81 Let $\mathcal{A}$ be a linear transformation on a finite dimensional vector space V and let V_1 and V_2 be subspaces of V. Answer true or false:

(a) $\mathcal{A}(V_1 \cap V_2) = \mathcal{A}(V_1) \cap \mathcal{A}(V_2)$.

(b) $\mathcal{A}(V_1 \cup V_2) = \mathcal{A}(V_1) \cup \mathcal{A}(V_2)$.

(c) $\mathcal{A}(V_1 + V_2) = \mathcal{A}(V_1) + \mathcal{A}(V_2)$.

(d) $\mathcal{A}(V_1 \oplus V_2) = \mathcal{A}(V_1) \oplus \mathcal{A}(V_2)$.

3.82 Let $\mathcal{A}$ be a linear transformation on a finite dimensional vector space V. Show that the following are equivalent:

(a) $\mathcal{A}$ has an inverse.

(b) V and $\operatorname{Im}\mathcal{A}$ have the same dimension.

(c) $\mathcal{A}$ maps a basis to a basis.

(d) The matrix representation of $\mathcal{A}$ under some basis is invertible.

(e) $\mathcal{A}$ is one-to-one; that is, $\operatorname{Ker}\mathcal{A} = \{0\}$.

(f) $\mathcal{A}$ is onto; that is, $\operatorname{Im}\mathcal{A} = V$.

What if V is infinite dimensional? What if $\mathcal{A}$ is a linear transformation from V to another vector space W?

3.83 Let $\mathcal{A}$ be a linear transformation on a vector space V, $\dim V = n$.

(a) If for some vector v, the vectors $v, \mathcal{A}(v), \mathcal{A}^2(v), \dots, \mathcal{A}^{n-1}(v)$ are linearly independent, show that every eigenvalue of $\mathcal{A}$ has only one corresponding eigenvector up to a scalar multiplication.

(b) If $\mathcal{A}$ has n distinct eigenvalues, show that there is a vector u such that $u, \mathcal{A}(u), \mathcal{A}^2(u), \dots, \mathcal{A}^{n-1}(u)$ are linearly independent.

3.84 Let $\mathcal{A}$ be a linear transformation on a vector space V, $\dim V = n$. If

$$\mathcal{A}^{n-1}(x) \neq 0, \quad \text{but} \quad \mathcal{A}^n(x) = 0, \quad \text{for some } x \in V,$$

show that

$$x, \mathcal{A}(x), \dots, \mathcal{A}^{n-1}(x)$$

are linearly independent, and thus form a basis of V. What are the eigenvalues of $\mathcal{A}$? Find the matrix representation of $\mathcal{A}$ under the basis.

3.85 Let $\mathcal{A}$ and $\mathcal{B}$ be linear transformations on a finite dimensional vector space. Show that if $\mathcal{AB} = \mathcal{I}$, then $\mathcal{BA} = \mathcal{I}$. Is this true for infinite dimensional vector spaces?

3.86 If $u_1, u_2, \dots, u_k$ are eigenvectors belonging to distinct eigenvalues $\lambda_1, \lambda_2, \dots, \lambda_k$ of a linear transformation, show that $u_1, u_2, \dots, u_k$ are linearly independent. Simply put, different eigenvectors belonging to distinct eigenvalues are linearly independent.

3.87 Let V and W be finite dimensional vector spaces, and let $\mathcal{A}$ be a linear transformation from V to W. Answer true or false:

(a) $\operatorname{Ker} \mathcal{A} = \{0\}$.

(b) If $\mathcal{A}(v) = 0$ only when $v = 0$, then $\dim V = \dim W$.

(c) If $\operatorname{Im} \mathcal{A} = \{0\}$, then $\mathcal{A} = 0$.

(d) If $V = W$ and $\operatorname{Im} \mathcal{A} \subseteq \operatorname{Ker} \mathcal{A}$, then $\mathcal{A} = 0$.

(e) If $V = W$ and $\operatorname{Im} \mathcal{A} \subseteq \operatorname{Ker} \mathcal{A}$, then $\mathcal{A}^2 = 0$.

(f) If $\dim V = \dim W$, then $\mathcal{A}$ is invertible.

(g) If $\dim V = \dim \operatorname{Im} \mathcal{A}$, then $\operatorname{Ker} \mathcal{A} = \{0\}$.

(h) $\operatorname{Ker} \mathcal{A}^2 \supseteq \operatorname{Ker} \mathcal{A}$.

(i) $\dim \operatorname{Ker} \mathcal{A} \leq \dim \operatorname{Im} \mathcal{A}$.

(j) $\dim \operatorname{Ker} \mathcal{A} \leq \dim V$.

(k) $\mathcal{A}$ is one-to-one if and only if $\operatorname{Ker} \mathcal{A} = \{0\}$.

(l) $\mathcal{A}$ is one-to-one if and only if $\dim V \leq \dim W$.

(m) $\mathcal{A}$ is onto if and only if $\operatorname{Im} \mathcal{A} = W$.

(n) $\mathcal{A}$ is onto if and only if $\dim V \geq \dim W$.

3.88 Let V and W be finite dimensional vector spaces, and let $\mathcal{A}$ be a linear transformation from V to W. Prove or disprove:

(a) If the vectors $\alpha_1, \alpha_2, \ldots, \alpha_n$ in V are linearly independent, then $\mathcal{A}\alpha_1, \mathcal{A}\alpha_2, \ldots, \mathcal{A}\alpha_n$ are linearly independent.

(b) If the vectors $\mathcal{A}\alpha_1, \mathcal{A}\alpha_2, \ldots, \mathcal{A}\alpha_n$ in W are linearly independent, then $\alpha_1, \alpha_2, \ldots, \alpha_n$ are linearly independent.

3.89 Let $\{\alpha_1, \alpha_2, \alpha_3\}$ be a basis for a three-dimensional vector space V. Let $\mathcal{A}$ be a linear transformation on V such that

$$\mathcal{A}(\alpha_1) = \alpha_1, \quad \mathcal{A}(\alpha_2) = \alpha_1 + \alpha_2, \quad \mathcal{A}(\alpha_3) = \alpha_1 + \alpha_2 + \alpha_3.$$

(a) Show that $\mathcal{A}$ is invertible.

(b) Find $\mathcal{A}^{-1}$.

(c) Find $2\mathcal{A} - \mathcal{A}^{-1}$.

3.90 Let $\mathcal{A}$ be the linear transformation defined on $\mathbb{R}^3$ by

$$\mathcal{A}(x, y, z) = (0, x, y).$$

Find the characteristic polynomials of $\mathcal{A}$, $\mathcal{A}^2$, $\mathcal{A}^3$.

3.91 If $\mathcal{A}$ is a linear transformation on $\mathbb{R}^3$ such that

$$\mathcal{A}\begin{pmatrix} 1 \\ 0 \\ 1 \end{pmatrix} = \begin{pmatrix} 2 \\ 3 \\ -1 \end{pmatrix}, \quad \mathcal{A}\begin{pmatrix} 1 \\ -1 \\ 1 \end{pmatrix} = \begin{pmatrix} 3 \\ 0 \\ -2 \end{pmatrix},$$

and

$$\mathcal{A}\begin{pmatrix} -2 \\ 7 \\ -1 \end{pmatrix} = \begin{pmatrix} 2 \\ 3 \\ -1 \end{pmatrix},$$

find $\operatorname{Im}\mathcal{A}$, a matrix representation of $\mathcal{A}$, and a formula for $\mathcal{A}(x)$.

3.92 Let $\{\epsilon_1, \epsilon_2, \epsilon_3, \epsilon_4\}$ be a basis for a vector space V of dimension 4, and let $\mathcal{A}$ be a linear transformation on V having matrix representation under the basis

$$A = \begin{pmatrix} 1 & 0 & 2 & 1 \\ -1 & 2 & 1 & 3 \\ 1 & 2 & 5 & 5 \\ 2 & -2 & 1 & -2 \end{pmatrix}.$$

(a) Find $\operatorname{Ker}\mathcal{A}$.

(b) Find $\operatorname{Im}\mathcal{A}$.

(c) Take a basis for $\operatorname{Ker}\mathcal{A}$, extend it to a basis of V, and then find the matrix representation of $\mathcal{A}$ under this basis.

3.93 Let $\mathcal{A}$ and $\mathcal{B}$ be linear transformations on $\mathbb{R}^2$. It is known that the matrix representation of $\mathcal{A}$ under the basis $\{\alpha_1 = (1,2),\ \alpha_2 = (2,1)\}$ is $\left(\begin{smallmatrix} 1 & 2 \\ 2 & 3 \end{smallmatrix}\right)$, and the matrix representation of $\mathcal{B}$ under the basis $\{\beta_1 = (1,1),\ \beta_2 = (1,2)\}$ is $\left(\begin{smallmatrix} 3 & 3 \\ 2 & 4 \end{smallmatrix}\right)$. Let $u = (3,3) \in \mathbb{R}^2$. Find

(a) The matrix of $\mathcal{A} + \mathcal{B}$ under β_1, β_2.

(b) The matrix of $\mathcal{AB}$ under α_1, α_2.

(c) The coordinate of $\mathcal{A}(u)$ under α_1, α_2.

(d) The coordinate of $\mathcal{B}(u)$ under β_1, β_2.

3.94 Let W be a subspace of a finite dimensional vector space V. If $\mathcal{A}$ is a linear transformation from W to V, show that $\mathcal{A}$ can be extended to a linear transformation on V?

3.95 Let W be an *invariant subspace* of a linear transformation $\mathcal{A}$ on a finite dimensional vector space V; that is, $\mathcal{A}(w) \in W$ for all $w \in W$.

(a) If $\mathcal{A}$ is invertible, show that W is also invariant under $\mathcal{A}^{-1}$.

(b) If $V = W \oplus W'$, is W' necessarily invariant under $\mathcal{A}$?

3.96 Let $\mathcal{A}$ be a linear transformation on $\mathbb{R}^2$ with the matrix $A = \left(\begin{smallmatrix} 2 & 1 \\ 0 & 2 \end{smallmatrix}\right)$ under the basis $\alpha_1 = (1, 0)$, $\alpha_2 = (0, 1)$. Let W_1 be the subspace of $\mathbb{R}^2$ spanned by α_1. Show that W_1 is invariant under $\mathcal{A}$ and that there does not exist a subspace W_2 invariant under $\mathcal{A}$ such that $\mathbb{R}^2 = W_1 \oplus W_2$.

3.97 Consider the vector space of all 2×2 real matrices. Let E_{ij} be the 2×2 matrix with (i, j)-entry 1 and other entries 0, i, $j = 1$, 2. Let

$$A = \begin{pmatrix} 1 & -1 \\ -1 & 1 \end{pmatrix}$$

and define

$$\mathcal{A}(u) = Au, \quad u \in M_2(\mathbb{R}).$$

(a) Show that $\mathcal{A}$ is a linear transformation on $M_2(\mathbb{R})$.

(b) Find the matrix of $\mathcal{A}$ under the basis E_{ij}, $i, j = 1, 2$.

(c) Find $\operatorname{Im} \mathcal{A}$, its dimension, and a basis.

(d) Find $\operatorname{Ker} A$, its dimension, and a basis.

3.98 A linear transformation $\mathcal{L}$ on a vector space V is said to be a *projector* if $\mathcal{L}^2 = \mathcal{L}$. Let $\mathcal{A}$ and $\mathcal{B}$ be projectors on the same vector space V. Show that $\mathcal{A}$ and $\mathcal{B}$ commute with $(\mathcal{A} - \mathcal{B})^2$; show also that

$$(\mathcal{A} - \mathcal{B})^2 + (\mathcal{I} - \mathcal{A} - \mathcal{B})^2 = \mathcal{I}.$$

3.99 Let $\{\epsilon_1, \epsilon_2, \epsilon_3, \epsilon_4\}$ be a basis for a vector space V of dimension 4. Define a linear transformation on V such that

$$\mathcal{A}(\epsilon_1) = \mathcal{A}(\epsilon_2) = \mathcal{A}(\epsilon_3) = \epsilon_1, \quad \mathcal{A}(\epsilon_4) = \epsilon_2.$$

Find $\operatorname{Ker} \mathcal{A}$, $\operatorname{Im} \mathcal{A}$, $\operatorname{Ker} \mathcal{A} + \operatorname{Im} \mathcal{A}$, and $\operatorname{Ker} \mathcal{A} \cap \operatorname{Im} \mathcal{A}$.

3.100 Define transformations $\mathcal{A}$ and $\mathcal{B}$ on $\mathbb{R}^2 = \{(x, y) \mid x, y \in \mathbb{R}\}$ by

$$\mathcal{A}(x, y) = (y, x)$$

and

$$\mathcal{B}(x, y) = (x - y, x - y).$$

(a) Show that $\mathcal{A}$ and $\mathcal{B}$ are linear transformations.

(b) Find the nontrivial invariant subspaces of $\mathcal{A}$.

(c) Find $\operatorname{Ker}\mathcal{B}$ and $\operatorname{Im}\mathcal{B}$.

(d) Show that $\dim(\operatorname{Ker}\mathcal{B}) + \dim(\operatorname{Im}\mathcal{B}) = 2$, but $\operatorname{Ker}\mathcal{B} + \operatorname{Im}\mathcal{B}$ is not a direct sum. Is the sum $\operatorname{Ker}\mathcal{B} + \operatorname{Im}\mathcal{B}^*$ a direct sum?

3.101 Define mappings $\mathcal{A}$ and $\mathcal{B}$ on the vector space $\mathbb{R}^n$ by

$$\mathcal{A}(x_1, x_2, \ldots, x_n) = (0, x_1, x_2, \ldots, x_{n-1})$$

and

$$\mathcal{B}(x_1, x_2, \ldots, x_n) = (x_n, x_1, x_2, \ldots, x_{n-1}).$$

(a) Show that $\mathcal{A}$ and $\mathcal{B}$ are linear transformations.

(b) Find $\mathcal{AB}$, $\mathcal{BA}$, $\mathcal{A}^n$, and $\mathcal{B}^n$.

(c) Find matrix representations of $\mathcal{A}$ and $\mathcal{B}$.

(d) Find dimensions of $\operatorname{Ker}\mathcal{A}$ and $\operatorname{Ker}\mathcal{B}$.

3.102 Let $\mathcal{A}$ be a linear transformation on an n-dimensional vector space V. If $\{\alpha_1, \ldots, \alpha_m\}$ is a basis for $\operatorname{Im}\mathcal{A}$ and if $\{\beta_1, \ldots, \beta_m\}$ is such a set of vectors of V that

$$\mathcal{A}(\beta_i) = \alpha_i, \quad i = 1, \ldots, m,$$

show that

$$V = \operatorname{Span}\{\beta_1, \ldots, \beta_m\} \oplus \operatorname{Ker}\mathcal{A}.$$

3.103 Let $\mathcal{A}$ be a linear transformation on a finite dimensional vector space V. Show that $\dim(\operatorname{Im}\mathcal{A}^2) = \dim(\operatorname{Im}\mathcal{A})$ if and only if

$$V = \operatorname{Im}\mathcal{A} \oplus \operatorname{Ker}\mathcal{A}.$$

Specifically, if $\mathcal{A}^2 = \mathcal{A}$, then $V = \operatorname{Im}\mathcal{A} \oplus \operatorname{Ker}\mathcal{A}$; is the converse true?

3.104 If $\mathcal{A}$ is an *idempotent* linear transformation on an n-dimensional vector space V; that is, $\mathcal{A}^2 = \mathcal{A}$, show that

(a) $\mathcal{I} - \mathcal{A}$ is idempotent.

(b) $(\mathcal{I} - \mathcal{A})(\mathcal{I} - t\mathcal{A}) = \mathcal{I} - \mathcal{A}$ for any scalar t.

(c) $(2\mathcal{A} - \mathcal{I})^2 = \mathcal{I}$.

(d) $\mathcal{A} + \mathcal{I}$ is invertible and find $(\mathcal{A} + \mathcal{I})^{-1}$.

(e) $\operatorname{Ker} \mathcal{A} = \{ x - \mathcal{A}x \mid x \in V \} = \operatorname{Im}(\mathcal{I} - \mathcal{A})$.

(f) $V = \operatorname{Im} \mathcal{A} \oplus \operatorname{Ker} \mathcal{A}$.

(g) $\mathcal{A}x = x$ for every $x \in \operatorname{Im} \mathcal{A}$.

(h) If $V = M \oplus L$, then there exists a unique linear transformation $\mathcal{B}$ such that $\mathcal{B}^2 = \mathcal{B}$, $\operatorname{Im} \mathcal{B} = M$, $\operatorname{Ker} \mathcal{B} = L$.

(i) Each eigenvalue of $\mathcal{A}$ is either 1 or 0.

(j) The matrix representation of $\mathcal{A}$ under some basis is

$$A = \operatorname{diag}(1, \ldots, 1, 0, \ldots, 0).$$

(k) $r(A) + r(A - I) = n$.

(l) $r(A) = \operatorname{tr} A = \dim(\operatorname{Im} \mathcal{A})$.

(m) $|A + I| = 2^{r(A)}$.

3.105 Let $\mathcal{A}$ and $\mathcal{B}$ be linear transformations on an n-dimensional vector space V over the complex field $\mathbb{C}$ satisfying $\mathcal{AB} = \mathcal{BA}$. Show that

(a) If λ is an eigenvalue of $\mathcal{A}$, then the eigenspace

$$V_\lambda = \{ x \in V \mid \mathcal{A}x = \lambda x \}$$

is invariant under $\mathcal{B}$.

(b) $\operatorname{Im} \mathcal{A}$ and $\operatorname{Ker} \mathcal{A}$ are invariant under $\mathcal{B}$.

(c) $\mathcal{A}$ and $\mathcal{B}$ have at least one common eigenvector (not necessarily belonging to the same eigenvalue).

(d) The matrix representations of $\mathcal{A}$ and $\mathcal{B}$ are both upper-triangular under some basis.

If $\mathbb{C}$ is replaced with $\mathbb{R}$, which of the above remain true?

3.106 Let $\mathcal{D}$ be the *differential operator* on $\mathbb{P}_n[x]$ over $\mathbb{R}$ defined as: if

$$p(x) = a_0 + a_1 x + a_2 x^2 + \cdots + a_{n-1}x^{n-1} \in \mathbb{P}_n[x],$$

then

$$\mathcal{D}(p(x)) = a_1 + 2a_2 x + \cdots + i a_i x^{i-1} + \cdots + (n-1)a_{n-1}x^{n-2}.$$

(a) Show that $\mathcal{D}$ is a linear transformation on $\mathbb{P}_n[x]$.

(b) Find the eigenvalues of $\mathcal{D}$ and $\mathcal{I} + \mathcal{D}$.

(c) Find the matrix representations of $\mathcal{D}$ under the bases $\{1, x, x^2, \ldots, x^{n-1}\}$ and $\{1, x, \frac{x^2}{2}, \ldots, \frac{x^{n-1}}{(n-1)!}\}$.

(d) Is a matrix representation of $\mathcal{D}$ diagonalizable?

3.107 Let $\mathcal{C}_\infty(\mathbb{R})$ be the vector space of real-valued functions on $\mathbb{R}$ having derivative of all orders.

(a) Consider the differential operator

$$\mathcal{D}_1(y) = y'' + ay' + by, \quad y \in \mathcal{C}_\infty(\mathbb{R}),$$

where a and b are real constants. Show that $y = e^{\lambda x}$ lies in $\operatorname{Ker}\mathcal{D}_1$ if and only if λ is a root of the quadratic equation

$$t^2 + at + b = 0.$$

(b) Consider the second differential operator

$$\mathcal{D}_2(y) = y'', \quad y \in \mathcal{C}_\infty(\mathbb{R}).$$

Show that $y = ce^{\lambda x}$ is an eigenvector of $\mathcal{D}_2$ for any constant $c \in \mathbb{R}$ and that every positive number is an eigenvalue of $\mathcal{D}_2$.

3.108 Consider $\mathbb{P}_n[x]$ over $\mathbb{R}$. Define

$$\mathcal{A}(p(x)) = xp'(x) - p(x), \quad p(x) \in \mathbb{P}_n[x].$$

(a) Show that $\mathcal{A}$ is a linear transformation on $\mathbb{P}_n[x]$.

(b) Find $\operatorname{Ker}\mathcal{A}$ and $\operatorname{Im}\mathcal{A}$.

(c) Show that $\mathbb{P}_n[x] = \operatorname{Ker}\mathcal{A} \oplus \operatorname{Im}\mathcal{A}$.

3.109 Let V be an n-dimensional vector space over $\mathbb{C}$ and $\mathcal{A}$ be a linear transformation with matrix representation under a basis $\{u_1, u_2, \dots, u_n\}$

$$A = \begin{pmatrix} \lambda & 0 & 0 & \dots & 0 & 0 \\ 1 & \lambda & 0 & \dots & 0 & 0 \\ 0 & 1 & \lambda & \dots & 0 & 0 \\ \vdots & \vdots & \vdots & \vdots & \vdots & \vdots \\ 0 & 0 & 0 & \dots & \lambda & 0 \\ 0 & 0 & 0 & \dots & 1 & \lambda \end{pmatrix};$$

that is,

$$\mathcal{A}(u_1, \dots, u_n) = (\mathcal{A}u_1, \dots, \mathcal{A}u_n) = (u_1, \dots, u_n)A.$$

Show that

(a) V is the only invariant subspace of $\mathcal{A}$ containing u_1.

(b) Any invariant subspace of $\mathcal{A}$ contains u_n.

(c) Each subspace

$$V_i = \operatorname{Span}\{u_{n-i+1}, \dots, u_n\}, \quad i = 1, 2, \dots, n$$

is invariant under $\mathcal{A}$, and $x \in V_i$ if and only if

$$(\mathcal{A} - \lambda\mathcal{I})^i x = 0.$$

(d) $V_1, V_2, \dots, V_n$ are the only invariant subspaces.

(e) $\operatorname{Span}\{u_n\}$ is the only eigenspace of $\mathcal{A}$.

(f) V cannot be written as a direct sum of two nontrivial invariant subspaces of $\mathcal{A}$.

Find an invertible matrix S such that $SAS^{-1} = A^t$.

3.110 Let $A \in M_n(\mathbb{C})$. Define a linear transformation $\mathcal{L}$ on $M_n(\mathbb{C})$ by

$$\mathcal{L}(X) = AX, \qquad X \in M_n(\mathbb{C}).$$

Show that $\mathcal{L}$ and A have the same set of eigenvalues. How are the characteristic polynomials of $\mathcal{L}$ and A related?

3.111 For the vector space $\mathbb{P}[x]$ over $\mathbb{R}$, define

$$\mathcal{A}f(x) = f'(x), \quad f(x) \in \mathbb{P}[x]$$

and

$$\mathcal{B}f(x) = xf(x), \quad f(x) \in \mathbb{P}[x].$$

Show that

(a) $\mathcal{A}$ and $\mathcal{B}$ are linear transformations.

(b) $\operatorname{Im} \mathcal{A} = \mathbb{P}[x]$ and $\operatorname{Ker} \mathcal{A} \neq \{0\}$.

(c) $\operatorname{Ker} \mathcal{B} = \{0\}$ and $\mathcal{B}$ does not have an inverse.

(d) $\mathcal{A}\mathcal{B} - \mathcal{B}\mathcal{A} = \mathcal{I}$.

(e) $\mathcal{A}^k\mathcal{B} - \mathcal{B}\mathcal{A}^k = k\mathcal{A}^{k-1}$ for every positive integer k.

3.112 Let V be a finite dimensional vector space and $\mathcal{A}$ be a linear transformation on V. Show that there exists a positive integer k so that

$$V = \operatorname{Im} \mathcal{A}^k \oplus \operatorname{Ker} \mathcal{A}^k.$$

Chapter 4

Special Matrices

Definitions and Facts

Hermitian Matrix. An n-square complex matrix $A = (a_{ij})$ is said to be a *Hermitian matrix* if $A^* = A$; that is, $a_{ij} = \overline{a_{ji}}$ for all i and j. In case of real matrices, we say that A is *real symmetric* if $A^t = A$. A square matrix A is *skew-Hermitian* if $A^* = -A$, equivalently $a_{ij} = -\overline{a_{ji}}$. It is immediate that the entries on the main diagonal of a Hermitian matrix are necessarily real. A matrix A is skew-Hermitian if and only if iA is Hermitian.

The eigenvalues of a Hermitian matrix are all real. Let λ be an eigenvalue of a Hermitian matrix A and $Ax = \lambda x$ for some nonzero vector x. Taking conjugate transpose yields $x^*A^* = \bar{\lambda}x^*$. It follows that

$$\lambda x^*x = x^*(\lambda x) = x^*Ax = x^*A^*x = \bar{\lambda}x^*x.$$

Since A is Hermitian, x^*Ax is real; because $x^*x > 0$, λ must be real.

Positive Semidefinite Matrices. An n-square complex matrix A is called a *positive semidefinite matrix* if $x^*Ax \geq 0$ for all $x \in \mathbb{C}^n$. And A is said to be *positive definite* if $x^*Ax > 0$ for all nonzero $x \in \mathbb{C}^n$.

A positive semidefinite matrix is necessarily Hermitian, all main diagonal entries are nonnegative, and so are the eigenvalues. The Hermity may be seen as follows. Since x^*Ax is real, $(x^*Ax)^* = x^*A^*x$. It follows that $x^*(A^* - A)x = 0$ for all $x \in \mathbb{C}^n$. Therefore all eigenvalues of $A^* - A$ are zero. Notice that $A^* - A$ is skew-Hermitian, thus diagonalizable. It is immediate that $A^* - A = 0$ and $A^* = A$. Another (direct) way to see this is to choose various vectors for x. Let x be the n-column vector with the p-th component 1, the q-th component $c \in \mathbb{C}$, and 0 elsewhere. Then

$$x^*Ax = a_{pp} + a_{qq}|c|^2 + a_{pq}c + a_{qp}\bar{c}.$$

Since $x^*Ax \geq 0$, setting $c = 0$ reveals $a_{pp} \geq 0$. Now that $a_{pq}c + a_{qp}\bar{c} \in \mathbb{R}$ for any $c \in \mathbb{C}$, putting $c = 1$, we see that a_{pq} and a_{qp} have the opposite imaginary parts. Talking $c = i$, we see that they have the same real part. Thus $a_{pq} = \overline{a_{qp}}$, namely, A is Hermitian. To see that the eigenvalues are nonnegative, let λ be an eigenvalue of A and $Ax = \lambda x$ for some nonzero vector x. Then $x^*Ax = \lambda x^*x \geq 0$. Therefore, $\lambda \geq 0$.

We write $A \geq 0$ ($A > 0$) to mean that A is a positive semidefinite (positive definite) matrix. For two Hermitian matrices A and B of the same size, we write $A \geq B$ or $B \leq A$ if $A - B \geq 0$.

Three important facts that we will use frequently:

(p_1) If $A \geq 0$, then $X^*AX \geq 0$ for all matrices X of appropriate sizes;

(p_2) If $A \geq 0$, then A has a unique positive semidefinite square root;

(p_3) If $A > 0$ and the block matrix $\begin{pmatrix} A & B \\ B^* & C \end{pmatrix} \geq 0$, then $C - B^*A^{-1}B \geq 0$.

Note that it is possible that for some real square matrix A, $x^tAx \geq 0$ for all real vectors x, but A is not positive semidefinite in the above sense. Take $A = \begin{pmatrix} 0 & 1 \\ -1 & 0 \end{pmatrix}$. It is easy to verify that $x^tAx = 0$ for all $x \in \mathbb{R}^2$.

Normal Matrices. An n-square complex matrix A is called a *normal matrix* if $A^*A = AA^*$; that is, A and A^* commute. Hermitian, skew-Hermitian, and positive semidefinite matrices are normal matrices.

Spectral Decomposition. Let A be an n-square complex matrix with (not necessarily different) complex eigenvalues $\lambda_1, \ldots, \lambda_n$. Then A is

1. Normal if and only if $A = U^* \operatorname{diag}(\lambda_1, \ldots, \lambda_n)U$ for some unitary matrix U, where $\lambda_1, \ldots, \lambda_n$ are *complex* numbers.

2. Hermitian if and only if $A = U^* \operatorname{diag}(\lambda_1, \ldots, \lambda_n)U$ for some unitary matrix U, where $\lambda_1, \ldots, \lambda_n$ are *real* numbers.

3. Positive semidefinite if and only if $A = U^* \operatorname{diag}(\lambda_1, \ldots, \lambda_n)U$ for some unitary matrix U, where $\lambda_1, \ldots, \lambda_n$ are *nonnegative* real numbers.

4. Positive definite if and only if $A = U^* \operatorname{diag}(\lambda_1, \ldots, \lambda_n)U$ for some unitary matrix U, where $\lambda_1, \ldots, \lambda_n$ are *positive* real numbers.

There are many more sorts of special matrices. For instance, the Hadamard matrix, Toeplitz matrix, stochastic matrix, nonnegative matrix, and M-matrix, etc. These matrices are useful in various fields. The Hermitian matrix, positive semidefinite matrix, and normal matrix are basic ones.

Chapter 4 Problems

4.1 Show that the following statements are equivalent:

(a) $A \in M_n(\mathbb{C})$ is *Hermitian*; that is, $A^* = A$.

(b) There is a unitary matrix U such that U^*AU is real diagonal.

(c) x^*Ax is real for every $x \in \mathbb{C}^n$.

(d) $A^2 = A^*A$.

(e) $A^2 = AA^*$.

(f) $\operatorname{tr} A^2 = \operatorname{tr}(A^*A)$.

(g) $\operatorname{tr} A^2 = \operatorname{tr}(AA^*)$.

Referring to (d), does A have to be Hermitian if $A^*(A^2) = A^*(A^*A)$?

4.2 Let A and B be $n \times n$ Hermitian matrices. Answer true or false:

(a) $A + B$ is Hermitian.

(b) cA is Hermitian for every scalar c.

(c) AB is Hermitian.

(d) ABA is Hermitian.

(e) If $AB = 0$, then $A = 0$ or $B = 0$.

(f) If $AB = 0$, then $BA = 0$.

(g) If $A^2 = 0$, then $A = 0$.

(h) If $A^2 = I$, then $A = \pm I$.

(i) If $A^3 = I$, then $A = I$.

(j) $-A$, A^t, $\bar{A}$, A^{-1} (if A is invertible) are all Hermitian.

(k) The main diagonal entries of A are all real.

(l) The eigenvalues of A are all real.

(m) The eigenvalues of AB are all real.

(n) The determinant $|A|$ is real.

(o) The trace $\operatorname{tr}(AB)$ is real.

(p) The eigenvalues of BAB are all real.

4.3 Let $A = (a_{ij}) \in M_n(\mathbb{C})$ have eigenvalues $\lambda_1, \lambda_2, \ldots, \lambda_n$. Show that

$$\sum_{i=1}^{n} \lambda_i^2 = \sum_{i,\, j=1}^{n} a_{ij} a_{ji}.$$

In particular, if A is Hermitian, then

$$\sum_{i=1}^{n} \lambda_i^2 = \sum_{i,\, j=1}^{n} |a_{ij}|^2.$$

4.4 Let A be an $n \times n$ Hermitian matrix. Let $\lambda_{\min}(A)$ and $\lambda_{\max}(A)$ be the smallest and largest eigenvalues of A, respectively. Denote

$$\|x\| = \sqrt{x^* x} \quad \text{for } x \in \mathbb{C}^n.$$

Show that

$$\lambda_{\min}(A) = \min_{\|x\|=1} x^* A x,$$

$$\lambda_{\max}(A) = \max_{\|x\|=1} x^* A x,$$

and for every unit vector $x \in \mathbb{C}^n$

$$\lambda_{\min}(A) \le x^* A x \le \lambda_{\max}(A).$$

Show that for Hermitian matrices A and B of the same size,

$$\lambda_{\max}(A) + \lambda_{\min}(B) \le \lambda_{\max}(A + B) \le \lambda_{\max}(A) + \lambda_{\max}(B).$$

4.5 Let $A \in M_n(\mathbb{C})$. Show that

(a) $x^* A x = 0$ for every $x \in \mathbb{R}^n$ if and only if $A^t = -A$.

(b) $x^* A y = 0$ for all x and y in $\mathbb{R}^n$ if and only if $A = 0$.

(c) $x^* A x = 0$ for every $x \in \mathbb{C}^n$ if and only if $A = 0$.

(d) $x^* A x$ is a fixed constant for all unit vectors $x \in \mathbb{C}^n$ if and only if A is a scalar matrix.

Does it follow that $A = B$ if A and B are $n \times n$ real matrices satisfying

$$x^* A x = x^* B x, \quad \text{for all } x \in \mathbb{R}^n?$$

4.6 Let A be an $n \times n$ Hermitian matrix with eigenvalues $\lambda_1, \lambda_2, \ldots, \lambda_n$. Show that

$$(A - \lambda_1 I)(A - \lambda_2 I) \cdots (A - \lambda_n I) = 0.$$

4.7 Let A be an $n \times n$ Hermitian matrix. If the determinant of A is negative; that is, $|A| < 0$, show that $x^*Ax < 0$ for some $x \in \mathbb{C}^n$; equivalently, if A is a positive semidefinite matrix, show that $|A| \geq 0$.

4.8 Let A and B be $n \times n$ Hermitian matrices. Show that $A + B$ is always Hermitian and that AB is Hermitian if and only if $AB = BA$.

4.9 Let A and B be Hermitian matrices of the same size. If AB is Hermitian, show that every eigenvalue λ of AB can be written as $\lambda = ab$, where a is an eigenvalue of A and b is an eigenvalue of B.

4.10 Let Y be a square matrix. A matrix X is said to be a k-th *root* of Y if $X^k = Y$. Let A, B, C, and D be, respectively, the following matrices

$$\begin{pmatrix} 1 & 2 & 1 \\ 2 & 4 & 2 \\ 1 & 2 & 1 \end{pmatrix}, \quad \begin{pmatrix} 0 & 1 & 0 \\ 0 & 0 & 0 \\ 0 & 0 & 0 \end{pmatrix}, \quad \begin{pmatrix} 0 & 1 \\ 0 & 0 \end{pmatrix}, \quad \begin{pmatrix} C & 0 \\ 0 & C \end{pmatrix}.$$

Show that

(a) A has a real symmetric cubic root.

(b) B does not have a complex cubic root.

(c) B has a square root.

(d) C does not have a square root.

(e) D has a square root.

(f) Every real symmetric matrix has a real symmetric k-th root.

(g) If $X^2 = Y$, then $|\lambda I - X|$ is a divisor of $|\lambda^2 I - Y|$.

Find a 2×2 matrix $X \neq I_2$ such that $X^3 = I_2$.

4.11 If A is an n-square invertible Hermitian matrix, show that A and A^{-1} are $*$-*congruent*; that is, $P^*AP = A^{-1}$ for some invertible matrix P.

4.12 Show that for any nonzero real number x, the matrix

$$A = \begin{pmatrix} 2 & 3x \\ \frac{3}{x} & 2 \end{pmatrix}$$

satisfies the equation $A^2 - 4A - 5I_2 = 0$. As a result, the equation has an infinite number of distinct 2×2 matrices as roots.

4.13 Let A be a 3×3 Hermitian matrix with eigenvalues $\lambda_1 < \lambda_2 < \lambda_3$. If a and b are the eigenvalues of some 2×2 principal submatrix of A, where $a < b$, show that $\lambda_1 \leq a \leq \lambda_2 \leq b \leq \lambda_3$.

4.14 Let A be a Hermitian matrix partitioned as $A = \left(\begin{smallmatrix} H & B \\ B^* & C \end{smallmatrix}\right)$. Show that

$$\lambda_{\min}(A) \leq \lambda_{\min}(H) \leq \lambda_{\max}(H) \leq \lambda_{\max}(A).$$

In particular, for each main diagonal entry a_{ii} of A

$$\lambda_{\min}(A) \leq a_{ii} \leq \lambda_{\max}(A).$$

4.15 Let A and B be n-square Hermitian matrices. Show that

(a) Neither $\operatorname{tr}(A^2) \leq (\operatorname{tr} A)^2$ nor $\operatorname{tr}(A^2) \geq (\operatorname{tr} A)^2$ holds in general.

(b) $\operatorname{tr}(AB)^k$ is real for every positive integer k.

(c) $\operatorname{tr}(AB)^2 \leq \operatorname{tr}(A^2B^2)$. Equality holds if and only if $AB = BA$.

(d) $[\operatorname{tr}(AB)]^2 \leq (\operatorname{tr} A^2)(\operatorname{tr} B^2)$. Equality holds if and only if one is a multiple of the other, i.e., $A = kB$ or $B = kA$ for a scalar k.

As the trace of AB is real, are the eigenvalues of AB all real?

4.16 Let A be an $n \times n$ Hermitian matrix with rank r. Show that all nonzero $r \times r$ principal minors of A have the same sign.

4.17 Let A be an $n \times n$ real symmetric matrix. Denote the sum of all entries of A by $S(A)$. Show that

$$S(A)/n \leq S(A^2)/S(A).$$

4.18 Let A be an $n \times n$ Hermitian matrix. Show that the following statements are equivalent:

(a) A is positive semidefinite; that is, $x^*Ax \geq 0$ for all $x \in \mathbb{C}^n$.

(b) All eigenvalues of A are nonnegative.

(c) $U^*AU = \operatorname{diag}(\lambda_1, \lambda_2, \ldots, \lambda_n)$ for some unitary matrix U, where λ_i's are all nonnegative.

(d) $A = B^*B$ for some matrix B.

(e) $A = T^*T$ for some $r \times n$ matrix T with rank $r = r(T) = r(A)$.

(f) All principal minors of A are nonnegative.

(g) $\operatorname{tr}(AX) \geq 0$ for all X positive semidefinite.

(h) $X^*AX \geq 0$ for all $n \times m$ matrix X.

4.19 Let A and B be $n \times n$ positive semidefinite matrices. Show that

$$A^2 + AB + BA + B^2$$

is always positive semidefinite. Construct an example showing that

$$A^2 + AB + BA, \quad \text{thus } AB + BA,$$

is not necessarily positive semidefinite in general. Prove that if A and $AB + BA$ are positive definite, then B is positive definite.

4.20 Let A and B be any two $m \times n$ matrices. Show that

$$A^*A + B^*B \geq \pm(A^*B + B^*A).$$

4.21 Let A and B be positive semidefinite matrices of the same size. If the largest eigenvalues of A and B are less than or equal to 1, show that

$$AB + BA \geq -\tfrac{1}{4}I.$$

4.22 Find the values of λ and μ so that the matrices are positive definite:

$$\begin{pmatrix} 1 & \lambda & -1 \\ \lambda & 4 & 2 \\ -1 & 2 & 4 \end{pmatrix}, \quad \begin{pmatrix} 1 & 1 & -1 \\ 1 & 2 & \mu \\ -1 & \mu & 3 \end{pmatrix}.$$

4.23 Let $A = (a_{ij})$ be an $n \times n$ Hermitian matrix such that the main diagonal entries of A are all equal to 1, i.e., all $a_{ii} = 1$. If A satisfies

$$\sum_{j=1}^{n} |a_{ij}| \le 2, \quad i = 1, 2, \ldots, n,$$

show that

(a) $A \ge 0$.

(b) $0 \le \lambda \le 2$, where λ is any eigenvalue of A.

(c) $0 \le \det A \le 1$.

4.24 Give an example of a non-Hermitian matrix all of whose principal minors and eigenvalues are nonnegative, the matrix, however, is not positive semidefinite.

4.25 Is it possible for some non-Hermitian matrix $A \in M_n(\mathbb{C})$ to satisfy $x^t Ax \ge 0$ for all $x \in \mathbb{R}^n$? $x^* Ax \ge 0$ for all $x \in \mathbb{C}^n$?

4.26 Let $A = (a_{ij})$ be an $n \times n$ positive semidefinite matrix. Show that

(a) $a_{ii} \ge 0$, $i = 1, 2, \ldots, n$, and if $a_{ii} = 0$, then the i-th row and i-th column of A consist entirely of 0.

(b) $a_{ii}a_{jj} \ge |a_{ij}|^2$ for each pair of i and j. In particular, the largest entry of A in absolute value is on the main diagonal.

(c) $\lambda_{\max} I - A \ge 0$, where $\lambda_{\max}$ is the largest eigenvalue of A.

(d) $|A| = 0$ if some principal submatrix of A is singular.

(e) There exists an $n \times n$ invertible matrix P such that

$$A = P^* \begin{pmatrix} I_{r(A)} & 0 \\ 0 & 0 \end{pmatrix} P.$$

Is it possible to choose a unitary matrix P?

(f) The transpose A^t and the conjugate $\bar{A}$ are positive semidefinite.

4.27 Let A be an $m \times n$ matrix and x be an n-column vector. Show that

$$(A^* A)x = 0 \quad \Leftrightarrow \quad Ax = 0$$

and

$$\operatorname{tr}(A^* A) = 0 \;\Leftrightarrow\; A^* A = 0 \;\Leftrightarrow\; A = 0.$$

4.28 Let A be $n \times n$ positive definite. Show that for every n-column vector x

$$x^* A^{-1} x = \max_y (x^* y + y^* x - y^* A y).$$

4.29 Let A and B be $n \times n$ positive semidefinite matrices. Show that

$$\operatorname{Im}(AB) \cap \operatorname{Ker}(AB) = \{0\}.$$

In particular, setting $B = I$,

$$\operatorname{Im} A \cap \operatorname{Ker} A = \{0\}.$$

4.30 Let $A \geq 0$; that is, A is positive semidefinite.

(a) Show that there is a unique matrix $B \geq 0$ such that $B^2 = A$. The matrix B is called the *square root* of A, denoted by $A^{\frac{1}{2}}$.

(b) Discuss the analog for normal matrices.

(c) Find $A^{\frac{1}{2}}$ when A is

$$\begin{pmatrix} 2 & 0 \\ 0 & 0 \end{pmatrix}, \quad \begin{pmatrix} 1 & 1 \\ 1 & 1 \end{pmatrix}, \quad \frac{1}{2}\begin{pmatrix} 5 & -3 \\ -3 & 5 \end{pmatrix}.$$

4.31 Let $A \in M_n(\mathbb{C})$ and C be a matrix commuting with A, i.e., $AC = CA$. Show that C commutes with A^2 and with $A^{\frac{1}{2}}$ when $A \geq 0$; that is,

$$A^2 C = C A^2 \quad \text{and} \quad A^{\frac{1}{2}} C = C A^{\frac{1}{2}} \text{ if } A \geq 0.$$

4.32 Let $A \in M_n(\mathbb{C})$. Show that

(a) If A is Hermitian, then $A^2 \geq 0$.

(b) If A is skew-Hermitian, then $-A^2 \geq 0$.

(c) If A is upper- (or lower-) triangular, then the eigenvalues and the main diagonal entries of A coincide.

Discuss the converse of each of (a), (b), and (c).

4.33 For $X \in M_n(\mathbb{C})$, define $f(X) = X^* X$. Show that f is a *convex function* on $M_n(\mathbb{C})$; that is, for any $t \in [0, 1]$, with $\tilde{t} = 1 - t$,

$$f(tA + \tilde{t}B) \leq t f(A) + \tilde{t} f(B), \quad A, B \in M_n(\mathbb{C}).$$

4.34 Let A and B be $n \times n$ positive semidefinite matrices. If the eigenvalues of A and B are all contained in the interval $[a, b]$, where $0 < a < b$, show that for any $t \in [0, 1]$, with $\tilde{t} = 1 - t$,

$$0 \leq tA^2 + \tilde{t}B^2 - (tA + \tilde{t}B)^2$$

and

$$tA^2 + \tilde{t}^2 B^2 - (tA + \tilde{t}B)^2 \leq \tfrac{1}{4}(a - b)^2 I.$$

4.35 Let A, $B \in M_n(\mathbb{C})$. Show that

(a) If $A > 0$ and B is Hermitian, then there exists an invertible matrix P such that $P^*AP = I$ and P^*BP is diagonal.

(b) If $A \geq 0$ and $B \geq 0$, then there exists an invertible matrix P such that both P^*AP and P^*BP are diagonal matrices. Can the condition $B \geq 0$ be weakened so that B is Hermitian?

(c) If $A > 0$ and $B \geq 0$, then

$$|A + B| \geq |A|.$$

Equality holds if and only if $B = 0$.

(d) If $A \geq 0$ and $B \geq 0$, then

$$|A + B| \geq |A| + |B|.$$

Equality holds if and only if $|A + B| = 0$ or $A = 0$ or $B = 0$.

(e) If $A \geq 0$ and $B \geq 0$, then

$$|A + B|^{\frac{1}{n}} \geq |A|^{\frac{1}{n}} + |B|^{\frac{1}{n}}.$$

(f) For $t \in [0, 1]$, $\tilde{t} = 1 - t$,

$$|A|^t |B|^{\tilde{t}} \leq |tA + \tilde{t}B|.$$

In particular, for every positive integer $k \leq n$,

$$\sqrt{|A||B|} \leq \frac{|A + B|}{2^k} \leq \frac{|A + B|}{2}.$$

(g) And also

$$\sqrt{|A||B|} \leq \frac{|A| + |B|}{2} \leq \frac{|A + B|}{2}.$$

4.36 Let A and B be positive definite matrices so that $A - B$ is positive semidefinite. Show that $|A| \geq |B|$ and that if $|A - \lambda B| = 0$ then $\lambda \geq 1$.

4.37 Let A, $B \in M_n(\mathbb{C})$ and $A \geq B \geq 0$. Show that

(a) $C^*AC \geq C^*BC$ for every $C \in M_{n\times m}(\mathbb{C})$.

(b) $A + C \geq B + D$, where $C \geq D$.

(c) $\operatorname{tr} A \geq \operatorname{tr} B$.

(d) $\lambda_{\max}(A) \geq \lambda_{\max}(B)$.

(e) $|A| \geq |B|$.

(f) $r(A) \geq r(B)$.

(g) $B^{-1} \geq A^{-1}$ (when the inverses exist).

(h) $A^{\frac{1}{2}} \geq B^{\frac{1}{2}}$. Does it follow that $A^2 \geq B^2$?

4.38 Let A be positive definite and B be Hermitian, both $n \times n$. Show that

(a) The eigenvalues of AB and $A^{-1}B$ are necessarily real.

(b) $A + B \geq 0$ if and only if $\lambda(A^{-1}B) \geq -1$, where $\lambda(A^{-1}B)$ denotes any eigenvalue of $A^{-1}B$.

(c) $r(AB)$, the rank of AB, equals the number of nonzero eigenvalues of AB. Is this true if $A \geq 0$ and B is Hermitian?

4.39 Let A and B be $n \times n$ Hermitian matrices.

(a) Give an example that the eigenvalues of AB are not real.

(b) If A or B is positive semidefinite, show that all the eigenvalues of AB are necessarily real.

(c) If A or B is positive definite, show that AB is diagonalizable.

(d) Give an example showing that the positive definiteness in (c) is necessary; that is, if one of A and B is positive semidefinite and the other is Hermitian, then AB need not be diagonalizable.

4.40 Let $\lambda_{\max}(X)$ and $\sigma_{\max}(X)$ denote, respectively, the largest eigenvalue and singular value of a square matrix X. For $A \in M_n(\mathbb{C})$, show that

$$\lambda_{\max}\left(\frac{A + A^*}{2}\right) \leq \sigma_{\max}(A) \quad \text{and} \quad \operatorname{tr}\left(\frac{A + A^*}{2}\right)^2 \leq \operatorname{tr}(A^*A).$$

4.41 Let A, $B \in M_n(\mathbb{C})$ be positive semidefinite. Show that

(a) $A^{\frac{1}{2}}BA^{\frac{1}{2}} \geq 0$.

(b) The eigenvalues of AB and BA are all nonnegative.

(c) AB is not necessarily positive semidefinite.

(d) AB is positive semidefinite if and only if $AB = BA$.

(e) $\operatorname{tr}(AB^2A) = \operatorname{tr}(BA^2B)$.

(f) $\operatorname{tr}(AB^2A)^{\frac{1}{2}} = \operatorname{tr}(BA^2B)^{\frac{1}{2}}$.

(g) $\operatorname{tr}(AB) \leq \operatorname{tr} A \operatorname{tr} B \leq \frac{1}{2}[(\operatorname{tr} A)^2 + (\operatorname{tr} B)^2]$.

(h) $\operatorname{tr}(AB) \leq \lambda_{\max}(A) \operatorname{tr} B$.

(i) $\operatorname{tr}(AB) \leq \frac{1}{4}(\operatorname{tr} A + \operatorname{tr} B)^2$.

(j) $\operatorname{tr}(AB) \leq \frac{1}{2}(\operatorname{tr} A^2 + \operatorname{tr} B^2)$.

Does it follow that $\operatorname{tr} A^{\frac{1}{2}} = \operatorname{tr} B^{\frac{1}{2}}$ if $\operatorname{tr} A = \operatorname{tr} B$?

4.42 Let A, B, C, and D be $n \times n$ positive semidefinite matrices.

(a) Show that $AB + BA$ is Hermitian.

(b) Is it true that $AB + BA \geq 0$?

(c) Is it true that $A^2 + B^2 \geq 2AB$?

(d) Is it true that $\operatorname{tr} A^2 + \operatorname{tr} B^2 \geq \operatorname{tr}(2AB)$?

(e) Is it true that $A^2 + B^2 \geq AB + BA$?

(f) Show that $\operatorname{tr}(AB) \leq \operatorname{tr}(CD)$ if $A \leq C$ and $B \leq D$.

(g) Show that $\lambda_{\max}(AB) \leq \lambda_{\max}(A)\lambda_{\max}(B)$.

(h) Show that for $t \in [0, 1]$ and $\tilde{t} = 1 - t$,

$$\lambda_{\max}(tA + \tilde{t}B) \leq t\lambda_{\max}(A) + \tilde{t}\lambda_{\max}(B).$$

In particular,

$$\lambda_{\max}(A + B) \leq \lambda_{\max}(A) + \lambda_{\max}(B).$$

(i) Discuss the analog of (g) for the case of three matrices.

4.43 Construct examples.

(a) Non-Hermitian matrices A and B have only positive eigenvalues, while AB has only negative eigenvalues.

(b) Is it possible that $A + B$ has only negative eigenvalues for matrices A and B having positive eigenvalues?

(c) Matrices A, B, and C are positive definite (thus their eigenvalues are all positive), while ABC has only negative eigenvalues.

(d) Can the matrices in (c) be 3×3 or any odd number size?

4.44 Let A, B be $n \times n$ matrices. If A is positive semidefinite, show that

$$A^2B = BA^2 \quad \text{if and only if} \quad AB = BA.$$

What if A is just Hermitian?

4.45 Let A, B, and C be $n \times n$ positive semidefinite matrices. If C commutes with AB and $A - B$, show that C commutes with A and B.

4.46 Let A be a positive definite matrix. If B is a square matrix such that $A - B^*AB$ is positive definite, show that $|\lambda| < 1$ for every eigenvalue λ of B. Is it true that $\sigma < 1$ for every singular value σ of B?

4.47 Let A, B, and C be complex matrices of appropriate sizes. Show that

$$\begin{pmatrix} A & B \\ B^* & C \end{pmatrix} > 0 \quad \Rightarrow \quad C - B^*A^{-1}B > 0.$$

4.48 Let A, B, C, and D be square matrices of the same size. Show that

$$\begin{pmatrix} A & B \\ B^* & C \end{pmatrix} \geq 0 \quad \Rightarrow \quad A + B + B^* + C \geq 0.$$

4.49 Let A, B, C, and D be n-square matrices. Prove or disprove that

$$\begin{pmatrix} A & B \\ B^* & C \end{pmatrix} \geq 0 \quad \Rightarrow \quad \begin{pmatrix} A & B^* \\ B & C \end{pmatrix} \geq 0.$$

4.50 Let A and B be n-square Hermitian matrices. Show that

$$\begin{pmatrix} A & B \\ B & A \end{pmatrix} \geq 0 \quad \Leftrightarrow \quad A \pm B \geq 0.$$

4.51 Let A and B be real square matrices of the same size. Show that

$$A + iB \geq 0 \;\Leftrightarrow\; A - iB \geq 0 \;\Leftrightarrow\; \begin{pmatrix} A & -B \\ B & A \end{pmatrix} \geq 0.$$

4.52 Let A be an n-square positive definite matrix with eigenvalues λ_1, λ_2, $\dots$, λ_n. Find the eigenvalues of the partitioned matrix

$$M = \begin{pmatrix} A & I \\ I & A^{-1} \end{pmatrix}.$$

4.53 Let A be an n-square complex matrix and let $M = \begin{pmatrix} 0 & A \\ A^* & 0 \end{pmatrix}$.

(a) Show that $\det M = (-1)^n |\det A|^2$.

(b) Find the eigenvalues of M.

(c) If $A \neq 0$, why is M never positive semidefinite?

(d) Find the eigenvalues of the matrix $N = \begin{pmatrix} I & A^* \\ A & I \end{pmatrix}$.

4.54 Recall that the *singular values* of a matrix X are defined to be the square roots of the eigenvalues of X^*X. Let $\sigma_{\max}(X)$ denote the largest singular value of the matrix X. For A, $B \in M_n(\mathbb{C})$, show that

$$\sigma_{\max}(AB) \leq \sigma_{\max}(A)\sigma_{\max}(B),$$

$$\sigma_{\max}(A + B) \leq \sigma_{\max}(A) + \sigma_{\max}(B),$$

$$\sigma_{\max}(A^2 - B^2) \leq \sigma_{\max}(A + B)\sigma_{\max}(A - B).$$

4.55 Find the singular values of the $n \times n$ real symmetric matrix

$$A = \begin{pmatrix} 1 & 1 & \dots & 1 \\ 1 & -1 & \dots & 0 \\ \vdots & \vdots & \ddots & \vdots \\ 1 & 0 & \dots & -1 \end{pmatrix}.$$

What are the eigenvalues of the matrix A?

4.56 Let $A = (a_{ij}) \in M_n(\mathbb{C})$ be a positive semidefinite matrix.

(a) Show that (the Hadamard determinantal inequality)

$$|A| \le a_{11}a_{22}\cdots a_{nn}.$$

Equality holds if and only if A is diagonal or some $a_{ii} = 0$.

(b) Write $A = \begin{pmatrix} B & C \\ C^* & D \end{pmatrix} \ge 0$, where B and D are square. Show that

$$|A| \le |B||D|.$$

Equality holds if and only if $C = 0$ or $|B| = 0$ or $|D| = 0$.

(c) With A partitioned as above in (b), where B, C, and D are square matrices of the same size, show that

$$\begin{pmatrix} |B| & |C| \\ |C^*| & |D| \end{pmatrix} \ge 0$$

or

$$|C^*C| \le |B||D|.$$

Equality holds if and only if B (or D) is singular or

$$D = C^*B^{-1}C.$$

What if B, C, and D are of different sizes?

(d) Show that for any $m \times n$ complex matrix $E = (e_{ij})$

$$|E^*E| \le \prod_{j=1}^{n}\sum_{i=1}^{m} |e_{ij}|^2.$$

(e) Let F be a complex matrix. If G is a submatrix consisting of some columns of F, and H is the submatrix consisting of the remaining columns of F, show that

$$|F^*F| \le |G^*G|\,|H^*H|.$$

(f) Show that for any square matrices X and Y of the same size,

$$|\det(X+Y)|^2 \le \det(I+XX^*)\det(I+Y^*Y).$$

4.57 Let $A \in M_n(\mathbb{C})$. Show that a necessary condition for $A^2 \ge 0$ is that all the eigenvalues of A are real. Is the converse true?

4.58 Let H be an $n \times n$ positive semidefinite matrix and write $H = A + iB$, where A and B are $n \times n$ real matrices. Show that

(a) A is positive semidefinite and $B^t = -B$.

(b) $a_{ss}a_{tt} \geq a_{st}^2 + b_{st}^2$ for each pair of s, t.

(c) $|H| \leq |A|$. When does equality hold?

(d) If A is singular, then H is singular.

Is the converse of (d) true? If the positive semidefiniteness of H is dropped, i.e., H is just Hermitian, which of the above remain true?

4.59 Let $A \in M_n(\mathbb{C})$. Show that

(a) A^*A and AA^* are unitarily similar.

(b) $A = HP$ for some $H \geq 0$ and P unitary.

(c) If A is an $m \times n$ matrix, then $A = AA^*Q$ for some matrix Q.

4.60 For any complex matrix A, the matrix $(A^*A)^{\frac{1}{2}}$ is called *modulus* of the matrix A, denoted by $\mathrm{m}(A)$. Let A be an $n \times n$ matrix, show that

(a) $\det(\mathrm{m}(A)) = |\det A|$.

(b) $A = \mathrm{m}(A)$ if $A \geq 0$.

(c) If $A = UDV$ is the singular value decomposition of A,

$$\mathrm{m}(A) = V^*DV \quad \text{and} \quad \mathrm{m}(A^*) = UDU^*.$$

(d) $\mathrm{m}(A)$ and $\mathrm{m}(A^*)$ are similar.

(e) $\mathrm{m}(A) = \mathrm{m}(A^*)$ if and only if A is normal.

(f) $\begin{pmatrix} \mathrm{m}(A) & A^* \\ A & \mathrm{m}(A^*) \end{pmatrix}$ is positive semidefinite.

(g) $\mathrm{m}(A)$ may not commute with A.

(h) $\mathrm{m}(A)H = H\mathrm{m}(A)$ if $AH = HA$ and H is Hermitian.

Find $\mathrm{m}(A)$ and $\mathrm{m}(A^*)$ for

$$\begin{pmatrix} 0 & 1 \\ 0 & 0 \end{pmatrix}, \quad \begin{pmatrix} 0 & 1 \\ 1 & 0 \end{pmatrix}, \quad \begin{pmatrix} 1 & 1 \\ 0 & 0 \end{pmatrix}.$$

4.61 Let A and B be both $m \times n$ complex matrices. Show that

$$\begin{pmatrix} A^*A & A^*B \\ B^*A & B^*B \end{pmatrix} \geq 0 \quad \text{and} \quad \begin{pmatrix} |A^*A| & |A^*B| \\ |B^*A| & |B^*B| \end{pmatrix} \geq 0.$$

Determine whether the following are true:

$$\begin{pmatrix} A^*A & B^*A \\ A^*B & B^*B \end{pmatrix} \geq 0 \quad \text{and} \quad \begin{pmatrix} |A^*A| & |B^*A| \\ |A^*B| & |B^*B| \end{pmatrix} \geq 0.$$

4.62 Let A be an $n \times n$ complex matrix with rank r. Show that

$$\frac{A + A^*}{2} = AA^*$$

if and only if $A = U \left(\begin{smallmatrix} I_r & 0 \\ 0 & 0 \end{smallmatrix}\right) U^*$ for some unitary matrix U.

4.63 Let $A \in M_{m \times n}(\mathbb{C})$ and $B \in M_{p \times n}(\mathbb{C})$. If $r(B) = p$, show that

$$AA^* \geq AB^*(BB^*)^{-1}BA^*.$$

4.64 Let A and B be $n \times n$ positive definite matrices. Show that

(a) If $|\lambda A - B| = 0$, then $\lambda > 0$.

(b) $|\lambda A - B| = 0$ has only solution $\lambda = 1$ if and only if $A = B$.

4.65 Let A and B be $m \times n$ matrices. Denote the *Hadamard* (or *Schur* or entrywise) *product* of A and B by $A \circ B$; that is, $A \circ B = (a_{ij}b_{ij})$.

Let A and B be $n \times n$ positive semidefinite matrices.

(a) Find $A \circ I$.

(b) Find $A \circ J$, where J is the $n \times n$ matrix of all entries 1.

(c) Show that $A \circ B \geq 0$.

(d) Show that $\lambda_{\max}(A \circ B) \leq \lambda_{\max}(A)\lambda_{\max}(B)$.

(e) Is it true that $A \circ B$ must be singular when A or B is singular?

(f) Show that $\operatorname{tr}(A \circ B) \leq \frac{1}{2} \operatorname{tr}(A \circ A + B \circ B)$.

4.66 Let A, $B \in M_{m\times n}(\mathbb{C})$. Let A_i and B_j denote, respectively, the i-th and the j-th column vectors of A and B, $i, j = 1, 2, \ldots, n$. Show that

$$(AA^*) \circ (BB^*) = (A \circ B)(A^* \circ B^*) + \sum_{i\neq j} (A_i \circ B_j)(A_i^* \circ B_j^*)$$

and

$$(A \circ B)(A^* \circ B^*) \leq (AA^*) \circ (BB^*).$$

In particular,

$$(A \circ B)^2 \leq A^2 \circ B^2, \quad \text{if } A, B \geq 0.$$

4.67 Let A and B be $n \times n$ *correlation matrices*, i.e., A and B are positive semidefinite and all entries on their main diagonals are 1. Prove

$$A^{\frac{1}{2}} \circ B^{\frac{1}{2}} \leq I.$$

4.68 Let $A \in M_n(\mathbb{C})$.

(a) If $A \geq 0$, show that $\begin{vmatrix} A & x \\ x^* & 0 \end{vmatrix} \leq 0$ for every column vector $x \in \mathbb{C}^n$. The inequality is strict if A is nonsingular and $x \neq 0$.

(b) If $A > 0$, find the inverse of $\begin{pmatrix} A & x \\ x^* & 0 \end{pmatrix}$.

4.69 Let A be a positive definite matrix. Partition A, A^{-1} conformably as

$$A = \begin{pmatrix} B & C \\ C^* & D \end{pmatrix}, \quad A^{-1} = \begin{pmatrix} U & V \\ V^* & W \end{pmatrix}.$$

(a) Show that U and W can be expressed, respectively, as

$$U = (B - CD^{-1}C^*)^{-1} = B^{-1} + B^{-1}CWC^*B^{-1},$$
$$W = (D - C^*B^{-1}C)^{-1} = D^{-1} + D^{-1}C^*UCD^{-1}.$$

(b) Show that

$$A - \begin{pmatrix} U^{-1} & 0 \\ 0 & 0 \end{pmatrix} \geq 0.$$

4.70 Let I be the $n \times n$ identity matrix. Find the eigenvalues of the matrix

$$G = \begin{pmatrix} I & I \\ I & I \end{pmatrix}.$$

4.71 Let $A > 0$. Show that

(a) $A + A^{-1} \geq 2I$.

(b) $A \circ A^{-1} \geq I$.

4.72 Let $A = (a_{ij})$ be an $n \times n$ matrix of nonnegative entries. If each row sum of A is equal to 1, namely, $\sum_{j=1}^{n} a_{ij} = 1$ for each i, show that

(a) For every eigenvalue λ of A, $|\lambda| \leq 1$.

(b) 1 is an eigenvalue of A.

(c) If A^{-1} exists, then each row sum of A^{-1} also equals 1.

4.73 Let A be a *real orthogonal matrix;* that is, A is real and $A^tA = AA^t = I$. Let $\lambda = a + ib$ be an eigenvalue of A and $u = x + iy$ be an eigenvector of λ, where a, b, x, y are real. If $b \neq 0$, show that

$$x^ty = 0 \quad \text{and} \quad x^tx = y^ty.$$

4.74 Let U be an $n \times n$ *unitary matrix*, i.e., $U^*U = UU^* = I$. Show that

(a) $U^* = U^{-1}$.

(b) U^t and $\overline{U}$ are unitary.

(c) UV is unitary for every $n \times n$ unitary matrix V.

(d) The eigenvalues of U are all equal to 1 in absolute value.

(e) If λ is an eigenvalue of U, then $\frac{1}{\lambda}$ is an eigenvalue of U^*.

(f) $|x^*Ux| \leq 1$ for every unit vector $x \in \mathbb{C}^n$.

(g) $\|Ux\| = 1$ for every unit vector $x \in \mathbb{C}^n$.

(h) Each row and column sum of $U \circ \overline{U} = (|u_{ij}|^2)$ equals 1.

(i) If x and y are eigenvectors of U belonging to distinct eigenvalues, then $x^*y = 0$.

(j) The columns (rows) of U form an orthonormal basis for $\mathbb{C}^n$.

(k) For any k rows of U, $1 \leq k \leq n$, there exist k columns such that the submatrix formed by the entries on the intersections of these rows and columns is nonsingular.

(l) $|\operatorname{tr}(UA)| \leq \operatorname{tr} A$ for every $n \times n$ matrix $A \geq 0$.

Which of the above statements imply that U is unitary?

4.75 For any complex matrix A, show that the following matrix is unitary

$$\begin{pmatrix} A & (I - AA^*)^{1/2} \\ (I - A^*A)^{1/2} & -A^* \end{pmatrix}.$$

4.76 Show that a square complex matrix U is unitary if and only if the column (row) vectors of U are all of length 1 and $|\det U| = 1$.

4.77 If the eigenvalues of $A \in M_n(\mathbb{C})$ are all equal to 1 in absolute value and if $\|Ax\| \leq 1$ for all unit vectors $x \in \mathbb{C}^n$, show that A is unitary.

4.78 Show that the $n \times n$ Vandermonde matrix U with the (i, j)-entry $\frac{1}{\sqrt{n}}\omega^{(i-1)(j-1)}$, where $\omega^n = 1$ and $\omega \neq 1$, is symmetric and unitary:

$$U = \frac{1}{\sqrt{n}} \begin{pmatrix} 1 & 1 & 1 & \cdots & 1 \\ 1 & \omega & \omega^2 & \cdots & \omega^{n-1} \\ 1 & \omega^2 & \omega^4 & \cdots & \omega^{2n-2} \\ \vdots & \vdots & \vdots & \vdots & \vdots \\ 1 & \omega^{n-1} & \omega^{2n-2} & \cdots & \omega^{(n-1)^2} \end{pmatrix}.$$

4.79 Let $A \in M_n(\mathbb{C})$ and let $U \in M_n(\mathbb{C})$ be a unitary matrix. Show that

$$\min_U \sigma_{\max}(U \circ A) \leq \frac{1}{\sqrt{n}} \Big(\sum_{i,\, j=1}^{n} |a_{ij}|^2 \Big)^{\frac{1}{2}}.$$

4.80 If A is an $n \times n$ real symmetric matrix, show that

$$I - iA \quad \text{and} \quad I + iA$$

are nonsingular and that

$$(I - iA)(I + iA)^{-1}$$

is unitary.

4.81 Let A be a nonsingular square complex matrix. Show that A is normal if and only if $A^{-1}A^*$ is unitary.

4.82 Find all 2×2 real orthogonal matrices.

4.83 Let $A = (a_{ij}) \in M_3(\mathbb{R})$ and $a_{11} \neq 0$. If the transpose A^t of A equals to the adjoint $\operatorname{adj}(A)$ of A, show that A is an orthogonal matrix.

4.84 Find an orthogonal matrix T such that $T^t AT$ is diagonal, where

$$A = \begin{pmatrix} 4 & 2 & 2 \\ 2 & 4 & 2 \\ 2 & 2 & 4 \end{pmatrix}.$$

4.85 Show that there do not exist real orthogonal matrices A and B satisfying $A^2 - B^2 = AB$. What if "orthogonal" is replaced by "invertible"?

4.86 If A and B are $n \times n$ real orthogonal matrices satisfying

$$|A| + |B| = 0,$$

show that

$$|A + B| = 0.$$

Can this be generalized to unitary matrices?

4.87 Let A and B be $n \times n$ real matrices. If $A > 0$ and $B = -B^t$; that is, A is positive definite and B is real skew-symmetric, show that

$$|A + B| > 0.$$

4.88 Let $A \in M_n(\mathbb{C})$. Show that if A is unitary, then so is the matrix

$$\frac{1}{\sqrt{2}} \begin{pmatrix} A & -A \\ A & A \end{pmatrix}.$$

4.89 Let $A \in M_n(\mathbb{C})$. Show that A can be written as $A = B + C$, where B is Hermitian and C is skew-Hermitian, and that A can also be written as $A = F + iG$, where both F and G are (unique) Hermitian.

4.90 Let A be a nonidentity square complex matrix.

(a) Can A be positive definite and unitary?

(b) Can A be Hermitian and unitary?

(c) Can A be upper-triangular (but not diagonal) and unitary?

4.91 Let $A = (a_{ij}) \in M_n(\mathbb{C})$ and have eigenvalues $\lambda_1, \lambda_2, \ldots, \lambda_n$. Show that the following statements are equivalent:

(a) A is *normal*; that is, $A^*A = AA^*$.

(b) $I - A$ is normal.

(c) A is unitarily diagonalizable; that is, there exists a unitary matrix U such that $U^*AU = \operatorname{diag}(\lambda_1, \lambda_2, \ldots, \lambda_n)$.

(d) There is a set of the unit eigenvectors of A that form an orthonormal basis for $\mathbb{C}^n$.

(e) Every eigenvector of A is an eigenvector of A^*.

(f) $A^* = AU$ for some unitary U.

(g) $A^* = VA$ for some unitary V.

(h) $\operatorname{tr}(A^*A) = \sum_{i,\, j=1}^n |a_{ij}|^2 = \sum_{i=1}^n |\lambda_i|^2$.

(i) The singular values of A are $|\lambda_1|, |\lambda_2|, \ldots, |\lambda_n|$.

(j) $\operatorname{tr}(A^*A)^2 = \operatorname{tr}[(A^*)^2A^2]$.

(k) $\|Ax\| = \|A^*x\|$ for every $x \in \mathbb{C}^n$.

(l) $A + A^*$ and $A - A^*$ commute.

(m) $A^*A - AA^*$ is positive semidefinite.

(n) A commutes with A^*A.

(o) A commutes with $AA^* - A^*A$.

4.92 Show that if A is a normal matrix, then A can be expressed as

(a) $A = B + iC$, where B and C are Hermitian and commute.

(b) $A = HP = PH$, where $H \geq 0$ and P is unitary.

Discuss the converses of (a) and (b).

4.93 If $A = \left(\begin{smallmatrix} B & C \\ 0 & D \end{smallmatrix}\right)$ is normal, what can be said about B, C, and D?

4.94 Let $A, B \in M_n(\mathbb{R})$. Show that if $M = A + iB$ is normal, Hermitian, orthogonal, or positive semidefinite, then so is the partitioned matrix

$$N = \begin{pmatrix} A & B \\ -B & A \end{pmatrix}.$$

4.95 Let $A \in M_n(\mathbb{C})$ be a normal matrix. Show that

(a) $\operatorname{Ker} A^* = \operatorname{Ker} A$.

(b) $\operatorname{Im} A^* = \operatorname{Im} A$.

(c) $\mathbb{C}^n = \operatorname{Im} A \oplus \operatorname{Ker} A$.

4.96 If A is a normal matrix and commutes with matrix B, show that the transpose conjugate A^* of A also commutes with B.

4.97 Let A be a normal matrix. Show that $A\bar{A} = 0 \Leftrightarrow AA^t = A^tA = 0$.

4.98 Let A and B be $n \times n$ normal matrices. Show that

(a) If $AB = BA$, then there exists a unitary matrix U such that U^*AU and U^*BU are both diagonal.

(b) If $AB = BA$, then AB is normal. What if $AB \neq BA$?

(c) If $AB^* = B^*A$, then both AB and BA are normal.

(d) $A + iB$ is normal if and only if $AB^* + A^*B$ is Hermitian.

(e) If AB is normal, then BA is normal.

Find two nonnormal matrices whose product is normal.

4.99 If A is a 3×3 matrix such that $A^2 = I$ and $A \neq \pm I$, show that the rank of one of $A + I$ and $A - I$ is 1 and the rank of the other is 2.

4.100 If A is a real matrix such that $A^3 + A = 0$, show that $\operatorname{tr} A = 0$.

4.101 Let A be an $n \times n$ matrix. If $A^k = I$ for some positive integer k, show that $\operatorname{tr}(A^{-1}) = \overline{\operatorname{tr}(A)}$. If such a k is less than n, show that A has repeated eigenvalues; that is, A cannot have n distinct eigenvalues.

4.102 Let A be an $n \times n$ real or complex matrix. If $A^k = I$ for some positive integer k, show that $T^{-1}AT$ is diagonal for some complex matrix T. For $B = \left(\begin{smallmatrix} 0 & -1 \\ 1 & 0 \end{smallmatrix}\right)$, show that $B^4 = I_2$ and there does not exist a real invertible matrix P such that $P^{-1}BP$ is diagonal.

4.103 A square matrix A is *nilpotent* if $A^k = 0$ for some positive integer k; *idempotent* or a *projection* if $A^2 = A$; *involutary* if $A^2 = I$. Show that

(a) A is nilpotent if and only if all eigenvalues of A are zero.

(b) A is idempotent if and only if A is similar to a diagonal matrix of the form $\operatorname{diag}(1, \dots, 1, 0, \dots, 0)$.

(c) A is involutary if and only if A is similar to a diagonal matrix of the form $\operatorname{diag}(1, \dots, 1, -1, \dots, -1)$.

4.104 Let A and B be nilpotent matrices of the same size. If $AB = BA$, show that $A + B$ and AB are also nilpotent.

4.105 If A is a nilpotent matrix, show that

(a) $I - A$ is invertible. Find $(I - A)^{-1}$.

(b) $I + A$ is also invertible.

(c) $\operatorname{tr} A = 0$.

(d) A is not diagonalizable when $A \neq 0$.

4.106 Let A and B be idempotent matrices of the same size. Show that $A + B$ is idempotent if and only if $AB = BA = 0$.

4.107 Let A and B be Hermitian. Show that if AB is idempotent, so is BA.

4.108 Let A be an n-square matrix of rank r, $r \geq 1$. Show that $A^2 = A$ if and only if there exist matrices B, $r \times n$, and C, $n \times r$, both of rank r, such that $A = BC$ and $CB = I_r$. Show further that if $A^2 = A$ then

$$|2I_n - A| = 2^{n-r} \quad \text{and} \quad |A + I_n| = 2^r.$$

4.109 Let A be an n-square matrix of rank r. If A satisfies $A^2 = A$ but is neither 0 nor I, show that for every positive integer k, $1 < k \leq n - r$, there exists a matrix B such that $AB = BA = 0$, and

$$(A + B)^{k+1} = (A + B)^k \neq (A + B)^{k-1}.$$

4.110 Let $A \in M_n(\mathbb{C})$ be a nonzero idempotent matrix; that is, $A^2 = A$.

(a) Find $|A+I|$ and $|A-I|$.

(b) Show that $r(A) = \operatorname{tr} A$.

(c) Show that $\dim(\operatorname{Im} A) = \operatorname{tr} A$.

4.111 Let $A, B \in M_n(\mathbb{C})$. Show that

(a) If $AB + BA = 0$ and if B is not nilpotent, then the matrix equation $AX + XA = B$ has no solution.

(b) If $A > 0$, then $AX + XA = B$ has a unique solution X. Moreover, if B is positive semidefinite, then so is X.

4.112 Let $A, B, C \in M_n(\mathbb{C})$. Show that the matrix equation $AX+XB = C$ has a unique solution if and only if $\begin{pmatrix} A & C \\ 0 & -B \end{pmatrix}$ and $\begin{pmatrix} A & 0 \\ 0 & -B \end{pmatrix}$ are similar.

4.113 A square complex matrix X is said to be *idempotent Hermitian* or called an *orthogonal projection* if $X^* = X$ and $X^2 = X$. Let A and B be $n \times n$ idempotent Hermitian matrices. Show that

$$B \le A \qquad \text{if and only if} \qquad AB = B.$$

4.114 Let A be an $n \times n$ idempotent Hermitian matrix. Show that

$$x \in \operatorname{Im} A \qquad \text{if and only if} \qquad x = Ax.$$

Let B be also an $n \times n$ idempotent Hermitian matrix. Show that

$$\operatorname{Im} A = \operatorname{Im} B \qquad \text{if and only if} \qquad A = B.$$

4.115 If $A \in M_n(\mathbb{C})$ is an involution, i.e., $A^2 = I$, show that the following assertions are equivalent:

(a) A is Hermitian.

(b) A is normal.

(c) A is unitary.

(d) All singular values of A are equal to 1.

4.116 Let A be an $n \times n$ involutary matrix, i.e., $A^2 = I$. Show that

(a) $X = \frac{1}{2}(I + A)$ and $Y = \frac{1}{2}(I - A)$ are idempotent, and $XY = 0$.

(b) $r(A + I) + r(A - I) = n$.

(c) A has only eigenvalues ± 1.

(d) $V = V_1 \oplus V_{-1}$, where V_1 and V_{-1} are the eigenspaces of the eigenvalues 1 and -1, respectively.

(e) $\operatorname{Im}(A - I) \subseteq \operatorname{Ker}(A + I)$.

Which of the above assertions imply $A^2 = I$?

4.117 Let A and B be $n \times n$ nonsingular matrices satisfying $ABA = B$ and $BAB = A$. Show that $M = A^2 = B^2$ is involutary; that is, $M^2 = I$.

4.118 Let A and B be n-square involutary matrices. Show that

$$\operatorname{Im}(AB - BA) = \operatorname{Im}(A - B) \cap \operatorname{Im}(A + B).$$

4.119 Let A, $B \in M_n(\mathbb{C})$ be such that $A = \frac{1}{2}(B + I)$. Show that A is idempotent, i.e., $A^2 = A$, if and only if B is an involution, i.e. $B^2 = I$.

4.120 Let A be a square matrix and λ be any nonzero scalar. Show that

$$\begin{pmatrix} A & A^* \\ A^* & A \end{pmatrix}, \quad \begin{pmatrix} A & \lambda^{-1}A \\ \lambda(I - A) & I - A \end{pmatrix}, \quad \begin{pmatrix} A & \lambda^{-1}A \\ -\lambda A & -A \end{pmatrix}$$

are normal, idempotent, and nilpotent matrices, respectively.

4.121 A *permutation matrix* is a matrix that has exactly one 1 in each row and each column, and all entries elsewhere are 0.

(a) How many $n \times n$ permutation matrices are there?

(b) Show that the product of two permutation matrices of the same size is also a permutation matrix. How about the sum?

(c) Show that any permutation matrix is invertible and its inverse is equal to its transpose.

(d) For what permutation matrices P, $P^2 = I$?

4.122 Let P be the $n \times n$ permutation matrix

$$P = \begin{pmatrix} 0 & 1 & 0 & \cdots & 0 \\ 0 & 0 & 1 & \cdots & 0 \\ 0 & 0 & 0 & \cdots & 0 \\ \vdots & \vdots & \vdots & & \vdots \\ 0 & 0 & 0 & \cdots & 1 \\ 1 & 0 & 0 & \cdots & 0 \end{pmatrix} = \begin{pmatrix} 0 & I_{n-1} \\ 1 & 0 \end{pmatrix}.$$

(a) Show that for any positive integer $k \leq n$,

$$P^k = \begin{pmatrix} 0 & I_{n-k} \\ I_k & 0 \end{pmatrix}$$

and

$$P^{n-1} = P^t, \quad P^n = I_n.$$

(b) Show that $P, P^2, \ldots, P^n$ are linearly independent.

(c) Show that $P^i + P^j$ is a normal matrix, $1 \leq i, j \leq n$.

(d) When is $P^i + P^j$ a symmetric matrix?

(e) Show that P is diagonalizable over $\mathbb{C}$ but not over $\mathbb{R}$ if $n \geq 3$.

(f) Show that for every P^i, where $1 < i < n$ and $(i, n) = 1$, there exists a permutation matrix T such that $T^{-1}P^iT = P$.

4.123 Let A be an invertible matrix with nonnegative integer entries. Show that if the sum of all entries of A^n is bounded for all n, then A is a permutation matrix. Equivalently, if the union over all n of the set of entries of A^n is finite, then A is a permutation matrix.

4.124 Let A be an $n \times n$ matrix so that every row and every column have one and only one nonzero entry that is either 1 or -1 and all other entries are 0. Show that $A^k = I$ for some positive integer k.

4.125 Let A be an $(n-1) \times n$ matrix of integers such that the row sums are all equal to zero. Show that $|AA^t| = nk^2$ for some integer k.

4.126 Let A be an $n \times n$ matrix all of whose entries are either 1 or -1 and whose rows are mutually orthogonal. Suppose A has an $s \times t$ submatrix whose entries are all 1. Show that $st \leq n$.

4.127 Let

$$A = \begin{pmatrix} 0 & 0 & 1 \\ 1 & 0 & 0 \\ 0 & -2 & 0 \end{pmatrix} \quad \text{and} \quad B = \begin{pmatrix} 0 & 1 & 0 \\ 0 & 0 & 1 \\ 1 & 0 & 0 \end{pmatrix}.$$

Show that

$$(A + tB)^3 = A^3 + t^3 B^3, \quad t \in \mathbb{R}.$$

4.128 Show that the matrix equation $X^4 + Y^4 = Z^4$ has nontrivial solutions:

$$\begin{pmatrix} 0 & x \\ 1 & 0 \end{pmatrix}^4 + \begin{pmatrix} 0 & y \\ 1 & 0 \end{pmatrix}^4 = \begin{pmatrix} 0 & z \\ 1 & 0 \end{pmatrix}^4,$$

where $x = 2pq$, $y = p^2 - q^2$, and $z = p^2 + q^2$, p and q are integers.

Chapter 5

Inner Product Spaces

Definitions and Facts

Inner Product Space. Let V be a vector space over the field $\mathbb{F}$, where $\mathbb{F}$ is $\mathbb{R}$ (or $\mathbb{C}$). An *inner product* on V is a mapping that assigns every ordered pair of vectors u and v in V a unique scalar in $\mathbb{F}$, denoted by $\langle u, v\rangle$. And the vector space V is called an *inner product space* (or *Euclidean space*) if the following are satisfied for all vectors $u, v, w \in V$ and scalar $\lambda \in \mathbb{F}$:

Positivity: $\langle u, u\rangle \geq 0$. Equality holds if and only if $u = 0$;

Homogeneity: $\langle \lambda u, v\rangle = \lambda\langle u, v\rangle$;

Linearity: $\langle u, v + w\rangle = \langle u, v\rangle + \langle u, w\rangle$;

Hermitian Symmetry: $\langle u, v\rangle = \overline{\langle v, u\rangle}$.

It is immediate that for all vectors u, v, and $w \in V$ and scalars λ and $\mu \in \mathbb{F}$,.

$$\langle u, \lambda v\rangle = \bar{\lambda}\langle u, v\rangle$$

and

$$\langle \lambda u + \mu v, w\rangle = \lambda\langle u, w\rangle + \mu\langle v, w\rangle.$$

When $\mathbb{F} = \mathbb{R}$, the inner product is also known as *dot product*, for which

$$\langle u, v\rangle = \langle v, u\rangle.$$

Let W be a subspace of the vector space V. If V is an inner product space, with the inner product being restricted to the vectors in W, then W is also an inner product space.

Examples.

- $\mathbb{R}^n$ is an inner product space with the standard inner product

$$\langle u, v\rangle = v^t u = u_1v_1 + u_2v_2 + \cdots + u_nv_n.$$

- $\mathbb{C}^n$ is an inner product space with the standard inner product

$$\langle u, v\rangle = v^* u = u_1\bar{v}_1 + u_2\bar{v}_2 + \cdots + u_n\bar{v}_n.$$

- Let l^2 be the vector space (over $\mathbb{C}$) of all infinite complex sequences $u = (u_1, u_2, \cdots)$ with the property that $\sum_{i=1}^{\infty} |u_i|^2 < \infty$. Then l^2 is an inner product space under the inner product

$$\langle u, v\rangle = \sum_{i=1}^{\infty} u_i\bar{v}_i.$$

- The vector space $C[a, b]$ of all continuous real-valued functions on the closed interval $[a, b]$ is an inner product space with the inner product

$$\langle f, g\rangle = \int_a^b f(t)g(t)dt.$$

- $M_{m\times n}(\mathbb{C})$ is an inner product space with the inner product

$$\langle A, B\rangle = \operatorname{tr}(B^*A).$$

Length or Norm of a Vector. Let u be a vector in an inner product space V. The nonnegative number $\sqrt{\langle u, u\rangle}$ is called the *length* or *norm* of the vector u and is denoted by $\|u\|$. The zero vector has length 0. A *unit vector* is a vector of norm 1. The norm has the following properties:

(l_1) $\|u\| \geq 0$ for all $u \in V$. Equality holds if and only if $u = 0$.

(l_2) $\|\lambda u\| = |\lambda|\, \|u\|$ for all $u \in V$ and $\lambda \in \mathbb{F}$.

(l_3) Triangle inequality: $\|u + v\| \leq \|u\| + \|v\|$ for all $u, v \in V$. Equality holds if and only if $v = 0$ or $u = 0$ or $u = \lambda v$ for some scalar $\lambda > 0$.

The *distance* between two vectors u and v is $d(u, v) = \|u - v\|$. One may show that for any two vectors u and v in an inner product space,

$$|\, \|u\| - \|v\|\, | \leq \|u - v\|$$

and

$$\|u+v\|^2 + \|u-v\|^2 = 2\|u\|^2 + 2\|v\|^2.$$

The Cauchy-Schwarz Inequality. Let u and v be any two vectors in an inner product space V. Then

$$|\langle u,\, v\rangle|^2 \le \langle u,\, u\rangle\, \langle v,\, v\rangle,$$

equivalently

$$|\langle u,\, v\rangle| \le \|u\|\, \|v\|.$$

Equality holds if and only if u and v are linearly dependent.

Orthogonal Vectors. Let u and v be vectors in an inner product space V. If the inner product $\langle u, v\rangle = 0$, then we say that u and v are *orthogonal* and write $u \perp v$. Nonzero orthogonal vectors are necessarily linearly independent. Suppose, say, $\lambda u + \mu v = 0$. Then $\lambda = 0$, thus $\mu = 0$, since

$$\langle \lambda u + \mu v, u\rangle = \lambda\langle u, u\rangle + \mu\langle v, u\rangle = \lambda\langle u, u\rangle = 0.$$

A subset S of V is called an *orthogonal set* if $u \perp v$ for all $u,\, v \in S$. S is further said to be an *orthonormal set* if S is an orthogonal set and all vectors in S are unit vectors. Two subsets S and T of V are said to be *orthogonal* if $u \perp v$ for all $u \in S$ and all $v \in T$.

We denote by $u^\perp$ and $S^\perp$ the collections of all vectors in V that are orthogonal to the vector u and subset S, respectively; that is,

$$u^\perp = \{\, v \in V \mid \langle u,\, v\rangle = 0 \,\}$$

and

$$S^\perp = \{\, v \in V \mid \langle u,\, v\rangle = 0 \quad \text{for all } u \in S \,\}.$$

These sets are called *orthogonal complements* of u and S, respectively. One may check that orthogonal complements are always subspaces of V. Obviously, $S \cap S^\perp = \{0\}$, $\{0\}^\perp = V$, and $V^\perp = \{0\}$. Moreover, $S \subseteq (S^\perp)^\perp$ and if S is a subspace of finite dimensional space, then $(S^\perp)^\perp = S$.

Orthogonal Basis; Orthonormal Basis. Let $\{\alpha_1, \alpha_2, \ldots, \alpha_n\}$ be a basis for an inner product space V. If $\alpha_1, \alpha_2, \ldots, \alpha_n$ are pairwise orthogonal; that is, $\langle \alpha_i, \alpha_j\rangle = 0$, whenever $i \neq j$, then we say that the basis is an *orthogonal basis.* If, in addition, every vector in the basis has length 1, we call such a

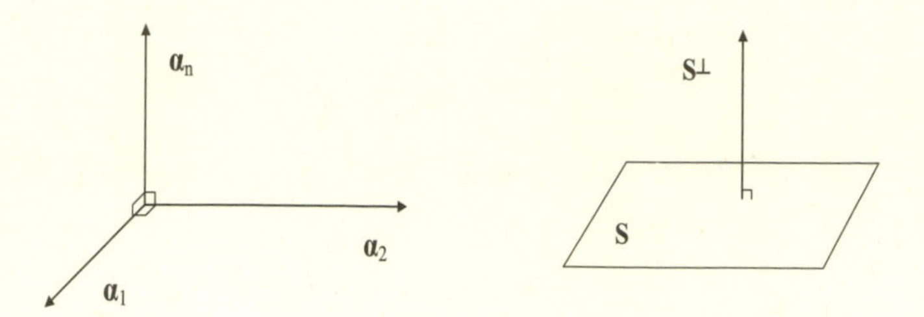

Figure 5.1: Orthogonality

basis an *orthonormal basis.* Thus $\alpha_1, \alpha_2, \dots, \alpha_n$ form an orthonormal basis for an n-dimensional inner product space if and only if

$$\langle \alpha_i, \alpha_j \rangle = \begin{cases} 0 & \text{if } i \neq j, \\ 1 & \text{if } i = j. \end{cases}$$

The standard basis $e_1, e_2, \dots, e_n$ are orthonormal basis for $\mathbb{R}^n$ and $\mathbb{C}^n$ under the usual inner product. The column (row) vectors of any $n \times n$ unitary matrix is also an orthonormal basis for $\mathbb{C}^n$.

If $\{u_1, u_2, \dots, u_k\}$ is an orthogonal set, then $\langle u_i, u_j \rangle = 0$ for all $i \neq j$. By putting $u = u_1 + u_2 + \cdots + u_k$ and computing $\|u\|$, we see that

$$\|u\|^2 = \left\| \sum_{i=1}^{k} u_i \right\|^2 = \sum_{i=1}^{k} \|u_i\|^2.$$

In particular, for two orthogonal vectors u and v,

$$\|u + v\|^2 = \|u\|^2 + \|v\|^2.$$

Let $\{\alpha_1, \alpha_2, \dots, \alpha_n\}$ be an orthonormal basis for an inner product space V. Then every vector u can be uniquely expressed as

$$u = \sum_{i=1}^{n} \langle u, \alpha_i \rangle \alpha_i.$$

The coordinates of u under the basis are $\langle u, \alpha_1 \rangle, \langle u, \alpha_2 \rangle, \dots, \langle u, \alpha_n \rangle$. And

$$\|u\|^2 = \sum_{i=1}^{n} |\langle u, \alpha_i \rangle|^2.$$

Chapter 5 Problems

5.1 Let V be an inner product space over $\mathbb{C}$. Show that for any nonzero vector $u \in V$, $\frac{1}{\|u\|}u$ is a unit vector and that for any $v, w \in V$,

$$\langle v, \langle v, w\rangle w\rangle = |\langle v, w\rangle|^2 = \langle v, w\rangle \langle w, v\rangle.$$

Is it true that

$$\langle \langle v, w\rangle v, w\rangle = |\langle v, w\rangle|^2?$$

5.2 Let V be an inner product space over $\mathbb{C}$. Show that for all vectors $u, v, w \in V$ and scalars $\lambda, \mu \in \mathbb{C}$ the following identities hold:

$$\langle \lambda u + \mu v, w\rangle = \lambda\langle u, w\rangle + \mu\langle v, w\rangle$$

and

$$\langle w, \lambda u + \mu v\rangle = \bar{\lambda}\langle w, u\rangle + \bar{\mu}\langle w, v\rangle.$$

5.3 Let V be an inner product space and $u, v, w \in V$. Is it true that

(a) $|\langle u, v\rangle| \le \|u\| + \|v\|$;

(b) $|\langle u, v\rangle| \le \frac{1}{2}(\|u\|^2 + \|v\|^2)$;

(c) $|\langle u, v\rangle| \le |\langle u, w\rangle| + |\langle w, v\rangle|$;

(d) $\|u + v\| \le \|u + w\| + \|w + v\|$; or

(e) $\|u + v\| \le \|u + w\| + \|w - v\|$?

5.4 Let $x = (x_1, x_2, \ldots, x_n) \in \mathbb{C}^n$ and $\|x\|_\infty = \max\{\, |x_1|, |x_2|, \ldots, |x_n|\,\}$. For $x, y \in \mathbb{C}^n$, define $\langle x, y\rangle_\infty = \|x\|_\infty \|y\|_\infty$. Check whether each of the following is satisfied:

(a) $\langle x, x\rangle_\infty \ge 0$. Equality holds if and only if $x = 0$;

(b) $\langle \lambda x, y\rangle_\infty = \lambda\langle x, y\rangle_\infty$;

(c) $\langle x, y + z\rangle_\infty = \langle x, y\rangle_\infty + \langle x, z\rangle_\infty$;

(d) $\langle x, y\rangle_\infty = \overline{\langle y, x\rangle_\infty}$.

5.5 For each pair of vectors x and y in $\mathbb{C}^3$, assign a scalar (x, y) as follows:

$$(x, y) = y^* \begin{pmatrix} 1 & 0 & 1 \\ 0 & 2 & 0 \\ 1 & 0 & 2 \end{pmatrix} x.$$

Show that $\mathbb{C}^3$ is an inner product space with respect to $(\cdot, \cdot)$. What if the entry 2 in the $(2,2)$-position is replaced by -2? That is, is $\mathbb{C}^3$ still an inner product space? Moreover, if x and y on the right-hand side are switched in the definition, is $\mathbb{C}^3$ still an inner product space?

5.6 Let

$$V = \left\{ x = \begin{pmatrix} x_1 \\ x_2 \\ x_3 \\ x_4 \end{pmatrix} \in \mathbb{R}^4 \;\middle|\; x_1 = x_3 + x_4,\ x_2 = x_3 - x_4 \right\}.$$

Show that V is a subspace of $\mathbb{R}^4$. Find a basis for V and for $V^\perp$.

5.7 Let $\alpha_1 = (1, 0, 0, 0)^t$ and $\alpha_2 = \left(0, \frac{1}{2}, \frac{1}{2}, \frac{1}{\sqrt{2}}\right)^t$. Find vectors α_3 and α_4 in $\mathbb{R}^4$ so that α_1, α_2, α_3, α_4 form an orthonormal basis for $\mathbb{R}^4$.

5.8 Define an inner product on $\mathbb{P}_4[x]$ over $\mathbb{R}$ as follows:

$$\langle f, g \rangle = \int_0^1 f(x)g(x)dx.$$

Let W be the subspace of $\mathbb{P}_4[x]$ consisting of polynomial 0 and all polynomials with degree 0; that is, $W = \mathbb{R}$. Find a basis for $W^\perp$.

5.9 Let $u_1, u_2, \dots, u_n \in \mathbb{C}^m$ be n column vectors of m components. Determine whether each of the following matrices is positive semidefinite:

$$\begin{pmatrix} u_1^*u_1 & u_1^*u_2 & \dots & u_1^*u_n \\ u_2^*u_1 & u_2^*u_2 & \dots & u_2^*u_n \\ \dots & \dots & \dots & \dots \\ u_n^*u_1 & u_n^*u_2 & \dots & u_n^*u_n \end{pmatrix}, \quad \begin{pmatrix} u_1^*u_1 & u_2^*u_1 & \dots & u_n^*u_1 \\ u_1^*u_2 & u_2^*u_2 & \dots & u_n^*u_2 \\ \dots & \dots & \dots & \dots \\ u_1^*u_n & u_2^*u_n & \dots & u_n^*u_n \end{pmatrix},$$

$$\begin{pmatrix} u_1u_1^* & u_1u_2^* & \dots & u_1u_n^* \\ u_2u_1^* & u_2u_2^* & \dots & u_2u_n^* \\ \dots & \dots & \dots & \dots \\ u_nu_1^* & u_nu_2^* & \dots & u_nu_n^* \end{pmatrix}, \quad \begin{pmatrix} u_1u_1^* & u_2u_1^* & \dots & u_nu_1^* \\ u_1u_2^* & u_2u_2^* & \dots & u_nu_2^* \\ \dots & \dots & \dots & \dots \\ u_1u_n^* & u_2u_n^* & \dots & u_nu_n^* \end{pmatrix}.$$

5.10 Let V be an inner product space and let $u_1, u_2, \dots, u_n$ be any n vectors in V. Show that the matrix, called *Gram matrix*,

$$G = \begin{pmatrix} \langle u_1, u_1 \rangle & \langle u_2, u_1 \rangle & \dots & \langle u_n, u_1 \rangle \\ \langle u_1, u_2 \rangle & \langle u_2, u_2 \rangle & \dots & \langle u_n, u_2 \rangle \\ \dots & \dots & \dots & \dots \\ \langle u_1, u_n \rangle & \langle u_2, u_n \rangle & \dots & \langle u_n, u_n \rangle \end{pmatrix}$$

is positive semidefinite. How about the following matrix H?

$$H = \begin{pmatrix} \langle u_1, u_1 \rangle & \langle u_1, u_2 \rangle & \dots & \langle u_1, u_n \rangle \\ \langle u_2, u_1 \rangle & \langle u_2, u_2 \rangle & \dots & \langle u_2, u_n \rangle \\ \dots & \dots & \dots & \dots \\ \langle u_n, u_1 \rangle & \langle u_n, u_2 \rangle & \dots & \langle u_n, u_n \rangle \end{pmatrix}.$$

5.11 Let $u, v \in \mathbb{C}^n$ have norms less than 1, i.e., $\|u\| < 1$, $\|v\| < 1$. Show that

$$\begin{pmatrix} \frac{1}{1-\langle u,u \rangle} & \frac{1}{1-\langle u,v \rangle} \\ \frac{1}{1-\langle v,u \rangle} & \frac{1}{1-\langle v,v \rangle} \end{pmatrix} \geq 0.$$

5.12 Show that a square matrix is positive semidefinite if and only if it is a Gram matrix. To be precise, $A \in M_n(\mathbb{C})$ is positive semidefinite if and only if there exist n vectors $u_1, u_2, \dots, u_n \in \mathbb{C}^n$ such that

$$A = (a_{ij}), \quad \text{where } a_{ij} = \langle u_j, u_i \rangle.$$

5.13 Show that $\langle \cdot, \cdot \rangle$ is an inner product for $\mathbb{C}^n$ if and only if there exists an n-square positive semidefinite matrix A such that for all $x, y \in \mathbb{C}^n$

$$\langle x, y \rangle = y^* A x.$$

5.14 Let V be a vector space over $\mathbb{C}$. A mapping $\langle \cdot, \cdot \rangle \colon V \times V \mapsto \mathbb{C}$ is called an *indefinite inner product* if for all vectors $u, v, w \in V$ and scalars $\lambda, \mu \in \mathbb{C}$, (i). $\langle \lambda u, v \rangle = \lambda \langle u, v \rangle$; (ii). $\langle u, v+w \rangle = \langle u, v \rangle + \langle u, w \rangle$; and (iii). $\langle u, v \rangle = \overline{\langle v, u \rangle}$. (I.e, the positivity condition in the definition of a regular inner product is removed.) Show that for any vectors u, v

$$\operatorname{Re}\langle u, v \rangle = \frac{1}{4}\big(\langle u+v, u+v \rangle - \langle u-v, u-v \rangle\big).$$

5.15 Let $x = (x_1, x_2, \ldots, x_n) \in \mathbb{C}^n$ and $\|x\|_\infty = \max\{ |x_1|, |x_2|, \ldots, |x_n| \}$. If $\|x + y\|_\infty = \|x\|_\infty + \|y\|_\infty$, must x and y be linearly dependent?

5.16 Let $\mathcal{A}$ be a linear transformation on an inner product space V. Show that for any unit vector $x \in V$

$$\langle \mathcal{A}(x),\, x\rangle\, \langle x,\, \mathcal{A}(x)\rangle \le \langle \mathcal{A}(x),\, \mathcal{A}(x)\rangle.$$

In particular, for $A \in M_n(\mathbb{C})$ and $x \in \mathbb{C}^n$ with $\|x\| = 1$,

$$|x^*Ax|^2 = (x^*A^*x)\,(x^*Ax) \le (x^*A^*)\,(Ax) = \|Ax\|^2$$

or

$$|x^*Ax| \le \|Ax\|,$$

with equality if and only if x is a unit eigenvector of A.

5.17 Let A be an $n \times n$ positive semidefinite matrix. Show that

$$\|(I - A)(I + A)^{-1}x\| \le \|x\|, \quad x \in \mathbb{C}^n.$$

Equality holds if and only if $x \in \operatorname{Ker} A$, or equivalently

$$(I - A)(I + A)^{-1}x = x.$$

5.18 Let V be an inner product space over $\mathbb{R}$.

(a) If $v_1, v_2, v_3, v_4 \in V$ are pairwise product negative; that is,

$$\langle v_i,\, v_j\rangle < 0, \quad i, j = 1, 2, 3, 4,\ i \ne j,$$

show that v_1, v_2, v_3 are linearly independent.

(b) Is it possible for four vectors in the xy-plane to have pairwise negative products? How about three vectors?

(c) Are v_1, v_2, v_3, v_4 in (a) necessarily linearly (in)dependent?

(d) Suppose that u, v, and w are three unit vectors in the xy-plane. What are the maximum and minimum values that

$$\langle u,\, v\rangle + \langle v,\, w\rangle + \langle w,\, u\rangle$$

can attain? and when?

5.19 For $\mathbb{P}_3[x]$ over $\mathbb{R}$, define the inner product as

$$\langle f, g\rangle = \int_{-1}^{1} f(x)g(x)dx.$$

(a) Is $f(x) = 1$ a unit vector in $\mathbb{P}_3[x]$?

(b) Find an orthonormal basis for the subspace $\text{Span}\{x, x^2\}$.

(c) Complete the basis in (b) to an orthonormal basis for $\mathbb{P}_3[x]$ with respect to the inner product.

(d) Show that $\langle f, g\rangle = 0$ for $f \in V_1$, $g \in V_2$, i.e., $V_1 \perp V_2$, where

$$V_1 = \text{Span}\{1, x\} \quad \text{and} \quad V_2 = \text{Span}\left\{x^2 - \frac{1}{3}, x^3 - \frac{3}{5}x\right\}.$$

(e) Show that $\mathbb{P}_3[x] = V_1 \oplus V_2$, where V_1, V_2 are defined as above.

(f) Is $[\cdot, \cdot]$ defined by

$$[f, g] = \int_{0}^{1} f(x)g(x)dx$$

also an inner product for $\mathbb{P}_3[x]$?

(g) Find a pair of vectors v and w in $\mathbb{P}_3[x]$ such that

$$\langle v, w\rangle = 0 \quad \text{but} \quad [v, w] \neq 0.$$

(h) Is the basis found in (c) an orthonormal basis for $\mathbb{P}_3[x]$ with respect to $[\cdot, \cdot]$?

5.20 Let $\{v_1, v_2, \ldots, v_n\}$ be an orthonormal basis for an inner product space V over $\mathbb{C}$. Show that for any $x \in V$,

$$x = \sum_{i=1}^{n} \langle x, v_i\rangle v_i$$

and

$$\langle x, x\rangle \geq \sum_{i=1}^{k} |\langle x, v_i\rangle|^2, \quad 1 \leq k \leq n.$$

Why are pairwise orthogonal nonzero vectors linearly independent?

5.21 For $M_n(\mathbb{C})$, the vector space of all $n \times n$ complex matrices over $\mathbb{C}$, define the inner product as

$$\langle A, B \rangle = \operatorname{tr}(B^*A), \quad A, B \in M_n(\mathbb{C}).$$

Show that

(a) $M_n(\mathbb{C})$ is an inner product space.

(b) $\operatorname{tr}(A^*A) = 0$ if and only if $A = 0$.

(c) $|\operatorname{tr}(AB)|^2 \leq \operatorname{tr}(A^*A)\operatorname{tr}(B^*B)$.

(d) $\operatorname{tr}(ABB^*A^*) \leq \operatorname{tr}(A^*A)\operatorname{tr}(B^*B)$ or $\|AB\| \leq \|A\| \, \|B\|$.

(e) $\|A^*A - AA^*\| \leq \sqrt{2}\, \|A\|^2$.

(f) If $\operatorname{tr}(AX) = 0$ for every $X \in M_n(\mathbb{C})$, then $A = 0$.

(g) $W = \{ X \in M_n(\mathbb{C}) \mid \operatorname{tr} X = 0 \}$ is a subspace of $M_n(\mathbb{C})$ and $W^\perp$ consists of scalar matrices; that is, if $\operatorname{tr}(AX) = 0$ for all $X \in M_n(\mathbb{C})$ with $\operatorname{tr} X = 0$, then $A = \lambda I$ for some scalar λ. Find the dimensions of W and $W^\perp$.

(h) If $\langle A, X \rangle \geq 0$ for all $X \geq 0$ in $M_n(\mathbb{C})$, then $A \geq 0$.

Is $M_n(\mathbb{C})$ an inner product space if the inner product is instead defined as $\langle A, B \rangle = \operatorname{tr}(A^*B)$, $\operatorname{tr}(AB^*)$, or $\operatorname{tr}(BA)$?

5.22 Consider $\mathbb{R}^2$ with the standard inner product. Are vectors $u = (1, 0)$ and $v = (1, -1)$ unit vectors? Are they mutually orthogonal? Find

(a) $u^\perp$, $v^\perp$.

(b) $u^\perp \cap v^\perp$.

(c) $\{u, v\}^\perp$.

(d) $(\operatorname{Span}\{u, v\})^\perp$.

(e) $\operatorname{Span}\{u^\perp, v^\perp\}$.

5.23 Find all 2×2 complex matrices that are orthogonal, in the sense of $\langle A, B \rangle = \operatorname{tr}(B^*A) = 0$, where $A, B \in M_2(\mathbb{C})$, to the matrices

$$\begin{pmatrix} 1 & 0 \\ 0 & -1 \end{pmatrix} \text{ and } \begin{pmatrix} 0 & 1 \\ 1 & 0 \end{pmatrix}.$$

5.24 For $A \in M_n(\mathbb{C})$, denote the *field of values* or *numerical range* of A by

$$F(A) = \{\, x^*Ax \mid x \in \mathbb{C}^n,\ \|x\| = 1 \,\}.$$

(a) Show that $F(A + cI) = F(A) + c$ for every $c \in \mathbb{C}$.

(b) Show that $F(cA) = cF(A)$ for every $c \in \mathbb{C}$.

(c) Why are the diagonal entries of A contained in $F(A)$?

(d) Show that the eigenvalues of A are contained in $F(A)$.

(e) Are the singular values of A contained in $F(A)$?

(f) Describe $F(A)$ when A is Hermitian.

(g) Describe $F(A)$ when A is positive semidefinite.

(h) Determine $F(A)$ when A is

$$\begin{pmatrix} 1 & 0 \\ 0 & 0 \end{pmatrix}, \quad \begin{pmatrix} 0 & 1 \\ 0 & 0 \end{pmatrix}, \quad \begin{pmatrix} 0 & 0 \\ 1 & 1 \end{pmatrix}, \quad \begin{pmatrix} 1 & 0 \\ 0 & 1+i \end{pmatrix},$$

and

$$\begin{pmatrix} 0 & 0 & 0 \\ 0 & 1 & 0 \\ 0 & 0 & 1+i \end{pmatrix}.$$

5.25 Let V be an inner product space.

(a) For any $x, y \in V$, show that

$$\|x + y\|^2 + \|x - y\|^2 = 2\|x\|^2 + 2\|y\|^2.$$

(b) Show that $\|x + y\| = \|x\| + \|y\|$ if and only if

$$\|sx + ty\| = s\|x\| + t\|y\|, \quad \text{for all } s, t \geq 0.$$

(c) If $\|x\| = \|y\|$, show that $x+y$ and $x-y$ are orthogonal. Explain this with a geometric graph.

(d) If x and y are orthogonal, show that

$$\|x + y\|^2 = \|x\|^2 + \|y\|^2.$$

(e) Is the converse of (d) true over $\mathbb{C}$? Over $\mathbb{R}$?

5.26 Let W be a subspace of an inner product space V and let $W^{\perp}$ be the *orthogonal complement* of W in V; that is,

$$W^{\perp} = \{\, x \in V \mid \langle x,\, y \rangle = 0 \ \text{ for all } \ y \in W \,\}.$$

Let $u \in V$. Show that $v \in W$ is a *projection* of u onto W; that is,

$$u = v + v', \quad \text{for some } v' \in W^{\perp},$$

if and only if

$$\|u - v\| \leq \|u - w\|, \quad \text{for every } w \in W.$$

5.27 Let W be any subspace of an inner product space V. Show that

$$V = W \oplus W^{\perp}.$$

Consequently,

$$\dim V = \dim W + \dim W^{\perp}.$$

5.28 If W is a subspace of an inner product space V, answer true or false:

(a) There is a unique subspace W' such that $W' + W = V$.

(b) There is a unique subspace W' such that $W' \oplus W = V$.

(c) There is a unique subspace W' such that $W' \oplus W = V$ and $\langle w, w' \rangle = 0$ for all $w \in W$ and $w' \in W'$.

5.29 If S is a subset of an inner product space V, answer true or false:

(a) $S \cap S^{\perp} = \{0\}$.

(b) $S \subseteq (S^{\perp})^{\perp}$.

(c) $(S^{\perp})^{\perp} = S$.

(d) $(S^{\perp})^{\perp} = \text{Span}(S)$.

(e) $[(S^{\perp})^{\perp}]^{\perp} = S^{\perp}$.

(f) $S^{\perp}$ is always a subspace of V.

(g) $S^{\perp} \oplus \text{Span}(S) = V$.

(h) $\dim S^{\perp} + \dim (S^{\perp})^{\perp} = \dim V$.

5.30 Let W_1 and W_2 be subspaces of an inner product space V. Show that

(a) $(W_1 + W_2)^\perp = W_1^\perp \cap W_2^\perp$.

(b) $(W_1 \cap W_2)^\perp = W_1^\perp + W_2^\perp$.

5.31 Let $S = \{u_1, u_2, \dots, u_p\}$ be an orthogonal set of nonzero vectors in an n-dimensional inner product space V; that is, $\langle u_i, u_j \rangle = 0$ if $i \neq j$. Let $v_1, v_2, \dots, v_q$ be vectors in V that are all orthogonal to S, namely, $\langle v_i, u_j \rangle = 0$ for all i and j. If $p + q > n$, show that the vectors $v_1, v_2, \dots, v_q$ are linearly independent.

5.32 Let A, B, and C be n-square complex matrices such that

$$\begin{pmatrix} A & B^* \\ B & C \end{pmatrix} \geq 0.$$

Show that

$$|\langle Bx, y \rangle|^2 \leq \langle Ax, x \rangle \, \langle Cy, y \rangle, \quad \text{for all } x, y \in \mathbb{C}^n.$$

In particular, for any $A \geq 0$,

$$|\langle Ax, y \rangle|^2 \leq \langle Ax, x \rangle \, \langle Ay, y \rangle, \quad \text{for all } x, y \in \mathbb{C}^n,$$

and for any $A > 0$,

$$|\langle x, y \rangle|^2 \leq \langle Ax, x \rangle \, \langle A^{-1}y, y \rangle, \quad \text{for all } x, y \in \mathbb{C}^n.$$

5.33 Let $\mathcal{A}$ be the linear transformation on an inner product space V:

$$\langle \mathcal{A}(u), v \rangle = -\langle u, \mathcal{A}(v) \rangle, \quad u, v \in V.$$

If λ is a real eigenvalue of $\mathcal{A}$, show that $\lambda = 0$.

5.34 Find the null space S for the following equation system, then find $S^\perp$:

$$\begin{aligned} x_1 - 2x_2 + 3x_3 - 4x_4 &= 0 \\ x_1 + 5x_2 + 3x_3 + 3x_4 &= 0. \end{aligned}$$

5.35 Let V_1 and V_2 be two subspaces of an inner product space V of finite dimension. If $\dim V_1 < \dim V_2$, show that there exists a nonzero vector in V_2 that is orthogonal to all vectors in V_1.

5.36 Let $\mathcal{A}$ be a *self-adjoint* linear operator (transformation) on an inner product space V of finite dimension over $\mathbb{C}$; that is,

$$\langle \mathcal{A}(u), v\rangle = \langle u, \mathcal{A}(v)\rangle, \quad \text{for all } u, v \in V.$$

Show that there exists an orthonormal basis for V in which every basis vector is an eigenvector of $\mathcal{A}$. In other words, there exists a set of eigenvectors of $\mathcal{A}$ that form an orthonormal basis for V.

5.37 Let $\mathcal{A}$ be a self-adjoint linear transformation on an inner product space V of finite dimension over $\mathbb{C}$, and let W be a k-dimensional subspace of V. If $\langle \mathcal{A}(x), x\rangle > 0$ for all nonzero vectors x in W, show that $\mathcal{A}$ has at least k positive eigenvalues (counting multiplicity).

5.38 If $\mathcal{A}$ and $\mathcal{B}$ are linear transformations from an inner product space V to an inner product space W such that

$$\langle \mathcal{A}(v), w\rangle = \langle \mathcal{B}(v), w\rangle, \quad \text{for all } v \in V \text{ and } w \in W,$$

show that $\mathcal{A} = \mathcal{B}$.

5.39 Let $\mathcal{A}$ be a linear operator on an inner product space V and $\mathcal{A}^*$ be the *adjoint* of $\mathcal{A}$, i.e., $\mathcal{A}^*$ is a linear transformation on V such that

$$\langle \mathcal{A}(x), y\rangle = \langle x, \mathcal{A}^*(y)\rangle, \quad \text{for all } x, y \in V.$$

Show that

(a) Such an $\mathcal{A}^*$ is unique.

(b) $(\mathcal{A}^*)^* = \mathcal{A}$.

(c) $\operatorname{Ker} \mathcal{A}^* = (\operatorname{Im} \mathcal{A})^{\perp}$.

(d) $\operatorname{Im} \mathcal{A}^* = (\operatorname{Ker} \mathcal{A})^{\perp}$.

(e) $V = \operatorname{Ker} \mathcal{A}^* \oplus \operatorname{Im} \mathcal{A} = \operatorname{Im} \mathcal{A}^* \oplus \operatorname{Ker} \mathcal{A}$.

(f) If the matrix of $\mathcal{A}$ under an orthonormal basis is A, then matrix of $\mathcal{A}^*$ under the same basis is A^*.

5.40 Let $\mathcal{A}$ be a linear transformation on a vector space V. Let $\langle \cdot, \cdot\rangle$ be an inner product on V. If one defines

$$[x, y] = \langle \mathcal{A}(x), \mathcal{A}(y)\rangle,$$

What $\mathcal{A}$ will make $[\cdot, \cdot]$ an inner product for V?

5.41 If $\mathcal{A}$ is a mapping on an inner product space V satisfying

$$\langle \mathcal{A}(x), \mathcal{A}(y) \rangle = \langle x, y \rangle, \quad \text{for all } x, y \in V,$$

show that $\mathcal{A}$ must be a linear transformation. Such an $\mathcal{A}$ is called an *orthogonal transformation.*

5.42 Let $\mathcal{A}$ be a linear operator on an inner product space V of dimension n. Show that the following statements are equivalent:

(a) $\mathcal{A}$ is orthogonal; that is, $\langle \mathcal{A}(u), \mathcal{A}(v) \rangle = \langle u, v \rangle$.

(b) $\|\mathcal{A}(u)\| = \|u\|$ for all vectors u.

(c) If $\{\alpha_1, \alpha_2, \dots, \alpha_n\}$ is an orthonormal basis for V, then so is $\{\mathcal{A}(\alpha_1), \mathcal{A}(\alpha_2), \dots, \mathcal{A}(\alpha_n)\}$.

(d) The matrix representation A of $\mathcal{A}$ under a basis is an orthogonal matrix; that is, $A^tA = AA^t = I$.

Can the condition (c) be replaced by (c$'$): If $\{\beta_1, \beta_2, \dots, \beta_n\}$ is an orthogonal basis for V, then so is $\{\mathcal{A}(\beta_1), \mathcal{A}(\beta_2), \dots, \mathcal{A}(\beta_n)\}$.

5.43 Let $\mathcal{A}$ be a linear transformation on an inner product space V of dimension n and let $\{\alpha_1, \alpha_2, \dots, \alpha_n\}$ be an orthogonal basis of V. If

$$\langle \mathcal{A}(\alpha_i), \mathcal{A}(\alpha_i) \rangle = \langle \alpha_i, \alpha_i \rangle, \quad i = 1, 2, \dots, n,$$

is $\mathcal{A}$ necessarily an orthogonal transformation?

5.44 Let V be an inner product space. As is known, a linear mapping $\mathcal{A}$ on V is an orthogonal (linear) transformation if and only if $\|\mathcal{A}(u)\| = \|u\|$ for all $u \in V$. Show that the word "linear" in the statement as a pre-condition is necessary; that is, show by an example that a mapping $\mathcal{L}$ on V satisfying $\|\mathcal{L}(u)\| = \|u\|$ for all $u \in V$ is not necessarily a linear transformation. Likewise, recall the distance of two vectors $u, v \in V$

$$d(u, v) = \|u - v\|.$$

If $\mathcal{D}$ is a mapping on V that preserves the distance of any two vectors

$$d(\mathcal{D}(u), \mathcal{D}(v)) = \|u - v\|,$$

show by example that $\mathcal{D}$ is not necessarily a linear transformation.

5.45 If $\{\alpha_1, \alpha_2, \ldots, \alpha_n\}$ and $\{\beta_1, \beta_2, \ldots, \beta_n\}$ are two sets of vectors of an inner product space V of dimension n. Does there always exist a linear transformation that maps each α_i to β_i? Show that if

$$\langle \alpha_i, \alpha_j \rangle = \langle \beta_i, \beta_j \rangle, \quad i, j = 1, 2, \ldots, n,$$

then there exists an orthogonal (linear) transformation $\mathcal{A}$ such that

$$\mathcal{A}(\alpha_i) = \beta_i, \quad i = 1, 2, \ldots, n.$$

5.46 If $\mathcal{A}$ and $\mathcal{B}$ are linear operators on an inner product space such that

$$\langle \mathcal{A}(u), \mathcal{A}(u) \rangle = \langle \mathcal{B}(u), \mathcal{B}(u) \rangle, \quad u \in V,$$

show that there exists an orthogonal operator $\mathcal{C}$ such that

$$\mathcal{A} = \mathcal{C}\mathcal{B}.$$

5.47 Let V be an inner product vector space over $\mathbb{F}$. A *linear functional* on V is a linear transformation from V to $\mathbb{F}$ and the *dual space* of V, denoted by V^*, is the vector space of all linear functionals on V.

(a) For $v \in V$, define a mapping $\mathcal{L}_v$ from V to $\mathbb{F}$ by

$$\mathcal{L}_v(u) = \langle u, v \rangle, \quad \text{for all } u \in V.$$

Show that $\mathcal{L}_v$ is a linear functional for every v.

(b) Let $\mathcal{L}$ be the mapping from V to V^* defined by

$$\mathcal{L}(v) = \mathcal{L}_v, \quad \text{for all } v \in V.$$

Show that $\mathcal{L}$ is linear.

(c) Show that $\mathcal{L}$ is one-to-one and onto.

(d) Find a basis for the vector space V^*.

5.48 Let $\mathcal{T}$ be an orthogonal transformation on an inner product space V. Show that $V = W_1 \oplus W_2$, where

$$W_1 = \{\, x \in V \mid \mathcal{T}(x) = x \,\} \quad \text{and} \quad W_2 = \{\, x - \mathcal{T}(x) \mid x \in V \,\}.$$

5.49 Let u be a unit vector in an n-dimensional inner product space V over $\mathbb{R}$. Define

$$\mathcal{A}(x) = x - 2\langle x, u\rangle u, \quad x \in V.$$

Show that

(a) $\mathcal{A}$ is an orthogonal transformation.

(b) If A is a matrix representation of $\mathcal{A}$, then $|A| = -1$.

(c) The matrix representation of $\mathcal{A}$ under any orthonormal basis is of the form $I - vv^t$, where v is some column vector.

(d) If $x = ku + y$ and $\langle u, y\rangle = 0$, then $\mathcal{A}(x) = -ku + y$.

(e) If $\mathcal{B}$ is an orthogonal transformation with 1 as an eigenvalue, and if the eigenspace of 1 is of dimension $n - 1$, then

$$\mathcal{B}(x) = x - 2\langle x, w\rangle w, \quad x \in V$$

for some unit vector $w \in V$.

5.50 Let V be an inner product space and W a nontrivial subspace of V.

(a) Find a linear transformation $\mathcal{P}$ on V, called *orthogonal projection* from V onto W, such that

$$\mathcal{P}(w) = w, \quad w \in W \text{ and } \mathcal{P}(w') = 0, \quad w' \in W^{\perp}.$$

(b) Show that

$$\mathcal{P}^2 = \mathcal{P}.$$

(c) Show that such a $\mathcal{P}$ is uniquely determined by W.

(d) Find a nonidentity linear transformation $\mathcal{P}'$ such that

$$\mathcal{P}(w) = \mathcal{P}'(w), \quad w \in W, \text{ but } \mathcal{P} \neq \mathcal{P}'.$$

(e) Show that for every $v \in V$, $\langle \mathcal{P}(v), v\rangle \geq 0$.

(f) Show that for every $v \in V$, $\|\mathcal{P}(v)\| \leq \|v\|$.

(g) Show that $\mathcal{I} - \mathcal{P}$ is the orthogonal projection onto $W^{\perp}$.

(h) Show that for every $v \in V$

$$\|v\|^2 = \|\mathcal{P}(v)\|^2 + \|(\mathcal{I} - \mathcal{P})(v)\|^2.$$

5.51 Let $\mathcal{P}_1$, $\mathcal{P}_2$, ..., $\mathcal{P}_m$ be *idempotent* linear transformations on an n-dimensional vector space V; that is,

$$\mathcal{P}_i^2 = \mathcal{P}_i, \quad i = 1, 2, \ldots, m.$$

(a) Show that if

$$\mathcal{P}_1 + \mathcal{P}_2 + \cdots + \mathcal{P}_m = \mathcal{I},$$

then

$$V = \operatorname{Im} \mathcal{P}_1 \oplus \operatorname{Im} \mathcal{P}_2 \oplus \cdots \oplus \operatorname{Im} \mathcal{P}_m$$

and

$$\mathcal{P}_i \mathcal{P}_j = 0, \;\; i, \; j = 1, 2, \ldots, m, \; i \neq j.$$

(b) Define an inner product for V such that each $\mathcal{P}_i$ is an orthogonal projection.

(c) Show that if

$$\mathcal{P}_i \mathcal{P}_j = 0, \;\; i, \; j = 1, 2, \ldots, m, \; i \neq j,$$

then

$$V = \operatorname{Im} \mathcal{P}_1 \oplus \operatorname{Im} \mathcal{P}_2 \oplus \cdots \oplus \operatorname{Im} \mathcal{P}_m \oplus \cap_{i=1}^{m} \operatorname{Ker} \mathcal{P}_i.$$

Hints and Answers for Chapter 1

1.1 (a) Yes. Dimension is 1. $\{1\}$ is a basis.

(b) Yes. Dimension is 2. $\{1, i\}$ is a basis.

(c) No, since $i \in \mathbb{C}$ and $1 \in \mathbb{R}$, but $i \cdot 1 = i \notin \mathbb{R}$.

(d) Yes. Dimension is infinite, since $1, \pi, \pi^2, \ldots$ are linearly independent over $\mathbb{Q}$.

(e) No, since $\sqrt{2} \in \mathbb{R}$ and $1 \in \mathbb{Q}$, but $\sqrt{2} \cdot 1 = \sqrt{2} \notin \mathbb{Q}$.

(f) No, since $\mathbb{Z}$ is not a field.

(g) Yes only over $\mathbb{Q}$, the dimension is 3, and $\{1, \sqrt{2}, \sqrt{5}\}$ is a basis.

1.2 (a) All vectors with initial point O and terminal points in the first quadrant.

(b) All vectors with initial point O and terminal points in the first or third quadrants.

1.3 Yes over $\mathbb{C}$, $\mathbb{R}$, and $\mathbb{Q}$. The dimensions are 2, 4, ∞, respectively.

1.4 Suppose that the vector space V has a nonzero element α. Then $\{r\alpha \mid r \in \mathbb{F}\}$ is an infinite set, where $\mathbb{F} = \mathbb{C}$, $\mathbb{R}$, or $\mathbb{Q}$.

If $u + v = 0$ and $w + v = 0$, then $u = u + (w + v) = (u + v) + w = w$.

1.5 For the first part, the addition and scalar multiplication for V are defined the same way as for $\mathbb{R}^2$. It is sufficient to notice that V is a line passing through O. If the scalar multiplication for V is defined to be $\lambda \odot (x, y) = (\lambda x, 0)$, then V is no longer a vector space since it is not closed under the scalar multiplication: $0 \neq 2(\lambda x)$ unless $\lambda x = 0$.

1.6 It is easy to check that $\mathbb{H}$ is closed under the usual matrix addition. As to scalar multiplication, if λ is a real number, then

$$\lambda \begin{pmatrix} a & b \\ -\bar{b} & \bar{a} \end{pmatrix} = \begin{pmatrix} \lambda a & \lambda b \\ -\overline{\lambda b} & \overline{\lambda a} \end{pmatrix} \in \mathbb{H}.$$

If λ is a nonreal complex number, then $\lambda \bar{a} \neq \overline{\lambda a}$. So $\mathbb{H}$ is not a vector space over $\mathbb{C}$ since it not closed under the scalar multiplication.

1.7 Check all the conditions for a vector space. For instance, the condition $(\lambda\mu)v = \lambda(\mu v)$ in the definition of vector space is satisfied, since

$$(ab) \boxdot x = x^{ab} = (x^b)^a = a \boxdot (b \boxdot x), \quad a,\ b \in \mathbb{R},\ x \in \mathbb{R}^+.$$

The dimension of the vector space is 1, since for any $x \in \mathbb{R}^+$,

$$x = (\log x) \boxdot 10.$$

Thus $\{10\}$ is a basis. Any two numbers in $\mathbb{R}^+$ are linearly dependent.

$\mathbb{R}^+$ is not a vector space over $\mathbb{R}$ with respect to $\boxtimes$ and $\boxplus$, since the condition $\lambda(u+v) = \lambda u + \lambda v$ in the definition is not satisfied:

$$2 = 2 \boxtimes (1 \boxplus 1) \neq (2 \boxtimes 1) \boxplus (2 \boxtimes 1) = 4.$$

1.8 It suffices to show that $\lambda_1\alpha_1, \lambda_2\alpha_2, \ldots, \lambda_n\alpha_n$ are linearly independent. Let $l_1, l_2, \ldots, l_n$ be scalars. If

$$0 = l_1(\lambda_1\alpha_1) + \cdots + l_n(\lambda_n\alpha_n) = (l_1\lambda_1)\alpha_1 + \cdots + (l_n\lambda_n)\alpha_n,$$

then each $l_i\lambda_i = 0$, thus $l_i = 0$, $i = 1, 2, \ldots, n$, since all $\lambda_i \neq 0$.

For $v = x_1\alpha_1 + \cdots + x_n\alpha_n = (x_1/\lambda_1)(\lambda_1\alpha_1) + \cdots + (x_n/\lambda_n)(\lambda_n\alpha_n)$, it follows that the coordinate of v under the basis $\{\lambda_1\alpha_1, \ldots, \lambda_n\alpha_n\}$ is $(x_1/\lambda_1, \ldots, x_n/\lambda_n)$. The coordinate of $w = \alpha_1 + \cdots + \alpha_n$ under $\{\alpha_1, \ldots, \alpha_n\}$ is $(1, \ldots, 1)$, under $\{\lambda_1\alpha_1, \ldots, \lambda_n\alpha_n\}$ is $(1/\lambda_1, \ldots, 1/\lambda_n)$.

1.9 (i) The vectors $v_1, v_2, \ldots, v_k$ form a basis of V if and only if they span V and they are linearly independent.

(ii) The vectors $v_1, v_2, \ldots, v_k$ form a basis of V if and only if every vector of V is a linear combination of these vectors and any vector in this set is not a linear combination of the remaining vectors.

1.10 (i) Yes for $k > n$; inconclusive for $k \leq n$.

(ii) No for $k < n$; inconclusive for $k \geq n$.

(iii) No in general.

1.11 Let $v = x_1\alpha_1 + x_2\alpha_2 + x_3\alpha_3$. Set $a_4 = -\frac{1}{4}(x_1 + x_2 + x_3)$ and $a_i = x_i + a_4$, $i = 1, 2, 3$. Then $v = a_1\alpha_1 + a_2\alpha_2 + a_3\alpha_3 + a_4\alpha_4$. Suppose $v = b_1\alpha_1 + b_2\alpha_2 + b_3\alpha_3 + b_4\alpha_4$ with $b_1 + b_2 + b_3 + b_4 = 0$. Since $\{\alpha_1, \alpha_2, \alpha_3\}$ is a basis, we have $b_1 - b_4 = x_1$, $b_2 - b_4 = x_2$, $b_3 - b_4 = x_3$, implying $-4b_4 = x_1 + x_2 + x_3$ and $b_4 = a_4$. Hence, $b_i = a_i$, $i = 1, 2, 3$.

For the case of n, if $\{\alpha_1, \ldots, \alpha_n\}$ is a basis for $\mathbb{R}^n$ and $\alpha_{n+1} = -(\alpha_1 + \cdots + \alpha_n)$, then every vector in $\mathbb{R}^n$ can be uniquely written as a linear combination of the vectors $\alpha_1, \ldots, \alpha_{n+1}$ with the sum of the coefficients equal to zero.

1.12 It is sufficient to show that 1, $(x-1)$, $(x-1)(x-2)$ are linearly independent. Let $\lambda_1 1 + \lambda_2(x-1) + \lambda_3(x-1)(x-2) = 0$. Then setting $x = 1$, $x = 2$, and $x = 3$, respectively, yields $\lambda_1 = \lambda_2 = \lambda_3 = 0$.

To see that W is a subspace of $\mathbb{P}_3[x]$, let p, $q \in W$. It follows that $(p+q)(1) = p(1) + q(1) = 0$. Thus $p + q \in W$. For any scalar λ, $(\lambda p)(1) = \lambda p(1) = 0$. So $\lambda p \in W$. Thus W is a subspace of $\mathbb{P}_3[x]$.

$\dim W = 2$, since $(x-1)$ and $(x-1)(x-2)$ form a basis of W.

1.13 (a) True.
(b) False.
(c) False.
(d) False.
(e) True.
(f) True.
(g) False.
(h) False.
(i) False.
(j) False.
(k) False.
(l) True.
(m) False.

1.14 (a) True.
(b) True.
(c) False.
(d) False.
(e) True.
(f) False.
(g) True.
(h) False.

(i) True.

(j) True.

1.15 Since $\alpha_3 = \alpha_1 + \alpha_2$, the vectors α_1, α_2, α_3 are linearly dependent. However, α_1 and α_2 are not proportional, so they are linearly independent and thus form a basis for $\text{Span}\{\alpha_1, \alpha_2, \alpha_3\}$. The dimension of the span is 2.

1.16 Since $\alpha_4 - \alpha_3 = \alpha_3 - \alpha_2 = \alpha_2 - \alpha_1$, we have $\alpha_3 = 2\alpha_2 - \alpha_1$ and $\alpha_4 = 3\alpha_2 - 2\alpha_1$. Obviously, α_1 and α_2 are linearly independently, and thus they form a basis for V and $\dim V = 2$.

1.17 (c) is true; others are false.

1.18 $k \neq 1$.

1.19 (i) Since α_1, α_2, and α_3 are linearly dependent, there are scalars x_1, x_2, x_3, not all zero, such that $x_1\alpha_1 + x_2\alpha_2 + x_2\alpha_3 = 0$. x_1 cannot be zero, otherwise α_2 and α_3 would be linearly dependent, which would contradict the linear independency of α_2, α_3, and α_4. It follows that $\alpha_1 = (-\frac{x_2}{x_1})\alpha_2 + (-\frac{x_3}{x_1})\alpha_3$, so α_1 is a linear combination of α_2 and α_3.

(ii) Suppose α_4 is a linear combination of α_1, α_2, and α_3. Let $\alpha_4 = y_1\alpha_1 + y_2\alpha_2 + y_3\alpha_3$. Substitute the α_1 as a linear combination of α_2 and α_3 in (i), we see that α_4 is a linear combination of α_2 and α_3. This is a contradiction to the linear independency of α_2, α_3, and α_4.

1.20 Let $x_1\alpha_1 + x_2\alpha_2 + x_3\alpha_3 = 0$. Then $x_1 + x_2 = 0$, $x_1 + x_3 = 0$, and $x_2 + x_3 = 0$. Thus $x_1 = x_2 = x_3 = 0$. The coordinates of u, v, and w under the basis are $(1, 1, -1)$, $\frac{1}{2}(1, 1, -1)$, and $\frac{1}{2}(1, 1, 1)$, respectively.

1.21 It is routine to check that W is closed under addition and scalar multiplication and that the given three matrices are linearly independent.

$$\begin{pmatrix} a & b \\ b & c \end{pmatrix} = a\begin{pmatrix} 1 & 0 \\ 0 & 0 \end{pmatrix} + b\begin{pmatrix} 0 & 1 \\ 1 & 0 \end{pmatrix} + c\begin{pmatrix} 0 & 0 \\ 0 & 1 \end{pmatrix}.$$

The coordinate is $(1, -2, 3)$.

1.22 (a) It's easy to verify that the conditions for a vector space are met.

(b) To show that $\{1, x, \ldots, x^{n-1}\}$ is a basis, let λ_0, λ_1, $\ldots$, λ_{n-1} be scalars such that $\lambda_0 + \lambda_1 x + \cdots + \lambda_{n-1}x^{n-1} = 0$. Setting $x = 0$ yields $\lambda_0 = 0$. In a similar way by factoring x each time, we see that $\lambda_1 = \cdots = \lambda_{n-1} = 0$. Thus $\{1, x, \ldots, x^{n-1}\}$ is a linearly independent set, thus, it is a basis. The one for $x - a$ is similar.

(c) $\left(f(a), \frac{f'(a)}{1!}, \ldots, \frac{f^{(n-1)}(a)}{(n-1)!}\right)$, where $f^{(i)}$ is the i-th derivative.

(d) It is sufficient to show that $f_i(x)$, $i = 1, \ldots, n$, are linearly independent. Let $k_1 f_1(x) + \cdots + k_n f_n(x) = 0$. Putting $x = a_i$ gives $k_i = 0$, $i = 1, \ldots, n$.

(e) If $f, g \in W$, then $f(1) = 0$ and $g(1) = 0$. Thus $(f+g)(1) = f(1) + g(1) = 0$; that is, $f + g \in W$. For $\lambda \in \mathbb{R}$, $(\lambda f)(1) = \lambda f(1) = 0$, so $\lambda f \in W$. It follows that W is a subspace.

(f) Yes, $\mathbb{P}[x]$ is a vector space. No, it is of infinite dimension.

(g) Obviously, $\mathbb{P}_n[x]$ is a proper subset of $\mathbb{P}[x]$. Then use (a).

1.23 Let $a \sin x + b \cos x = 0$. Taking $x = 0$ gives $b = 0$; putting $x = \frac{\pi}{2}$ yields $a = 0$. So $\sin x$ and $\cos x$ are linearly independent. In the same way, we see that $\sin^2 x$ and $\cos^2 x$ are linearly independent.

For $y = a \sin x + b \cos x$, it is easy to check that $y'' = -y$.

1, $\sin^2 x$, and $\cos^2 x$ are linearly dependent since $\sin^2 x + \cos^2 x = 1$.

$\operatorname{Span}\{\sin x, \cos x\} \cap \mathbb{R} = \{0\}$ and $\operatorname{Span}\{\sin^2 x, \cos^2 x\} \cap \mathbb{R} = \mathbb{R}$.

1.24 The vectors are linearly independent if and only if $t \neq 1$.

1.25 Linearly independent vectors of S are also linearly independent vectors of V. This gives (a). If $\dim S = \dim V$, then a basis of S is also a basis of V, so $S = V$ and (b) holds. To see (c), if $\alpha = \{\alpha_1, \alpha_2, \ldots, \alpha_k\}$ is a basis of S, and if every $v \in V$ is a linear combination of the vectors in α, then α is a basis for V by definition. Otherwise, there exists a vector $\beta \in V$ such that $\alpha_1, \alpha_2, \ldots, \alpha_k, \beta$ are linearly independent. Inductively, α can be extended to a basis of V. For (d), one may take V to be the xy-plane and S to be the line $y = x$.

1.26 (a) $\{e_1, e_2, \ldots, e_n\}$ and $\{\epsilon_1, \epsilon_2, \ldots, \epsilon_n\}$ are linearly independent sets. Over $\mathbb{C}$, yes. Over $\mathbb{R}$, no, since the dimension of $\mathbb{C}^n$ over $\mathbb{R}$ is $2n$.

(b) $$A = \begin{pmatrix} 1 & -1 & 0 & \cdots & 0 & 0 \\ 0 & 1 & -1 & \cdots & 0 & 0 \\ 0 & 0 & 1 & \cdots & 0 & 0 \\ \vdots & \vdots & \vdots & & \vdots & \vdots \\ 0 & 0 & 0 & \cdots & 1 & -1 \\ 0 & 0 & 0 & \cdots & 0 & 1 \end{pmatrix}.$$

(c) $B = (\epsilon_1, \epsilon_2, \ldots, \epsilon_n)$.

(d) $(-1, \ldots, -1, n)$.

(e) Since the dimension of $\mathbb{R}^n$ over $\mathbb{R}$ is n.

(f) $e_1, e_2, \ldots, e_n$ and $(i, 0, \ldots, 0)^t$, where $i = \sqrt{-1}$.

1.27 Let $l_1\alpha_1 + l_2(\alpha_1 + \alpha_2) + \cdots + l_n(\alpha_1 + \alpha_2 + \cdots + \alpha_n) = 0$. Then

$$(l_1 + l_2 + \cdots + l_n)\alpha_1 + (l_2 + \cdots + l_n)\alpha_2 + \cdots + l_n\alpha_n = 0.$$

Since $\alpha_1, \alpha_2, \ldots, \alpha_n$ are linearly independent, the coefficient of α_n is l_n, thus $l_n = 0$. The coefficient of α_{n-1} is $l_{n-1} + l_n$, so $l_{n-1} = 0$. Inductively, $l_1 = l_2 = \cdots = l_{n-3} = 0$.

Let $x_1(\alpha_1 + \alpha_2) + x_2(\alpha_2 + \alpha_3) + \cdots + x_n(\alpha_n + \alpha_1) = 0$. Then

$$(x_1 + x_n)\alpha_1 + (x_1 + x_2)\alpha_2 + \cdots + (x_{n-1} + x_n)\alpha_n = 0$$

and

$$x_1 + x_n = 0,\ x_1 + x_2 = 0,\ \ldots,\ x_{n-1} + x_n = 0.$$

The system of these equations has a nonzero solution if and only if

$$\begin{vmatrix} 1 & 0 & 0 & \cdots & 0 & 1 \\ 1 & 1 & 0 & \cdots & 0 & 0 \\ 0 & 1 & 1 & \cdots & 0 & 0 \\ \vdots & \vdots & \vdots & & \vdots & \vdots \\ 0 & 0 & 0 & \cdots & 1 & 0 \\ 0 & 0 & 0 & \cdots & 1 & 1 \end{vmatrix} = 1 + (-1)^{n+1} = 0.$$

If n is even, the vectors are linearly dependent. If n is odd, they are linearly independent and thus form a basis. The converse is also true.

1.28 Let $x_1\alpha_1 + x_2\alpha_2 + x_3\alpha_3 = 0$. By solving the system of linear equations, we can get $x_1 = x_2 = x_3 = 0$. So α is a basis for $\mathbb{R}^3$. Similarly β is also a basis. [Note: The easiest way to see that α or β is a basis is to show that the determinant $\det(\alpha_1,\ \alpha_2,\ \alpha_3) \neq 0$. See Chapter 2.]

If A is a matrix such that $\beta = \alpha A$, then $A = \alpha^{-1}\beta$. This gives

$$A = \begin{pmatrix} 2 & 3 & 4 \\ 0 & -1 & 0 \\ -1 & 0 & -1 \end{pmatrix}.$$

If the coordinate of u under α is $(2, 0, -1)$ (the first column of A), then

$$u = (\alpha_1, \alpha_2, \alpha_3)(2, 0, -1)^t = \beta A^{-1}(2, 0, -1)^t = \beta(1, 0, 0)^t.$$

That is, the coordinate of u under the basis β is $(1, 0, 0)$.

1.29 Let $a_1\alpha_1 + \cdots + a_n\alpha_n - b\beta = 0$, where $a_1, a_2, \ldots, a_n, b$ are not all zero. We claim $b \neq 0$. Otherwise, $a_1 = a_2 = \cdots = a_n = 0$, since $\alpha_1, \alpha_2, \ldots, \alpha_n$ are linearly independent. It follows that

$$\beta = \frac{a_1}{b}\alpha_1 + \cdots + \frac{a_n}{b}\alpha_n.$$

To see the uniqueness, let $\beta = c_1\alpha_1 + \cdots + c_n\alpha_n$. Then

$$\left(\frac{a_1}{b} - c_1\right)\alpha_1 + \cdots + \left(\frac{a_n}{b} - c_n\right)\alpha_n = 0$$

and, since $\alpha_1, \ldots, \alpha_n$ are linearly independent, we have

$$\frac{a_i}{b} - c_i = 0 \quad \text{or} \quad c_i = \frac{a_i}{b}, \quad i = 1, \ldots, n.$$

1.30 If $\alpha_1, \alpha_2, \ldots, \alpha_n$ are linearly dependent, let $a_1\alpha_1 + \cdots + a_n\alpha_n = 0$, where not all a's are zero. Let k be the largest index such that $a_k \neq 0$ and $a_i = 0$ when $i > k$. Then a_k will do. The other way is obvious.

1.31 (a) It is obvious that $V \times W$ is closed under the addition and scalar multiplication. If 0_v and 0_w are zero vectors of V and W, respectively, then $(0_v, 0_w)$ is the zero vector of $V \times W$. It is routine to check that other conditions for a vector space are also satisfied.

(b) If $\{\alpha_1, \alpha_2, \ldots, \alpha_m\}$ is a basis for V and $\{\beta_1, \beta_2, \ldots, \beta_n\}$ is a basis for W, one may show that (α_i, β_j), $i = 1, 2, \ldots, m$, $j = 1, 2, \ldots, n$, form a basis for $V \times W$.

(c) mn.

(d) Identify $(x, (y, z)) \in \mathbb{R} \times \mathbb{R}^2$ with $(x, y, z) \in \mathbb{R}^3$.

(e) Let $e_1 = (1, 0)$, $e_2 = (0, 1)$. Then e_1, e_2 are a basis for $\mathbb{R}^2$. Let E_{ij} be the 2×2 matrix with (i, j)-entry 1 and all other entries 0, $i, j = 1, 2$. Then E_{11}, E_{12}, E_{21}, E_{22} are a basis for $M_2(\mathbb{R})$. The eight vectors (e_s, E_{ij}), $s = 1, 2$, $i, j = 1, 2$, form a basis for $\mathbb{R}^2 \times M_2(\mathbb{R})$.

(f) 16.

1.32 (a) True.

(b) True.

(c) True.

(d) True. The converse is also true.

(e) False.

(f) False.

(g) False.

(h) False. Take $r = 1$, $(1, 0)$, $(-1, 0)$, $(0, 1)$, and $(0, -1)$ in $\mathbb{R}^2$.

1.33 If $s + t \geq n$, we have nothing to prove. Without loss of generality, let $\alpha_1, \dots, \alpha_s$ be a basis for U and $\beta_1, \dots, \beta_t$ be a basis for V. Then every vector $\alpha_i + \beta_j$, thus every vector in W, is a linear combination of $\alpha_1, \dots, \alpha_s, \beta_1, \dots, \beta_t$. It follows that $\dim W \leq s + t$.

1.34 Suppose $a_0(tu + v) + a_1\alpha_1 + \cdots + a_r\alpha_r = 0$ for scalars $a_0, a_1, \dots, a_r$. We first claim $a_0 = 0$. Otherwise, dividing both ides by a_0, since u is a linear combination of $\alpha_1, \dots, \alpha_r$, we see that v is a linear combination of $\alpha_1, \dots, \alpha_r$, a contradiction. Now that $a_0 = 0$, the linear independence of $\alpha_1, \dots, \alpha_r$ implies $a_1 = a_2 = \cdots = a_r = 0$.

1.35 If X, $Y \in V$, then $AX = XA$ and $AY = YA$. Thus $A(X + Y) = AX + AY = XA + YA = (X + Y)A$; that is, $X + Y \in V$. For any scalar k, $A(kX) = k(AX) = k(XA) = (kX)A$, so $kX \in V$. Therefore, V is closed under the matrix addition and scalar multiplication. Namely, V is a vector space. For the given A, $\dim V = 5$ and the matrices that commute with A take the form

$$\begin{pmatrix} a & b & 0 \\ c & d & 0 \\ -3a - c - e & -3b - d + e & e \end{pmatrix}.$$

1.36 Let E_{st} denote the $n \times n$ matrix with the (s, t)-entry 1 and 0 elsewhere.

(a) E_{st}, $1 \leq s$, $t \leq n$, form a basis. Dimension is n^2.

(b) E_{st}, iE_{st}, $1 \leq s$, $t \leq n$, form a basis. Dimension is $2n^2$.

(c) E_{st}, $1 \leq s$, $t \leq n$, form a basis. Dimension is n^2.

(d) $E_{st} + E_{ts}$, $s \leq t$, $i(E_{st} - E_{ts})$, $s < t$, form a basis. Dimension is n^2.

(e) $E_{st} + E_{ts}$, $s \leq t$, form a basis. Dimension is $\frac{n(n+1)}{2}$.

(f) $E_{st} - E_{ts}$, $s < t$, $i(E_{st} + E_{ts})$, $s \leq t$, form a basis. Dimension is n^2.

(g) $E_{st} - E_{ts}$, $s < t$, form a basis. Dimension is $\frac{n(n-1)}{2}$.

(h) E_{st}, $1 \leq s \leq t \leq n$, form a basis. Dimension is $\frac{n(n+1)}{2}$.

(i) E_{st}, $1 \leq t \leq s \leq n$, form a basis. Dimension is $\frac{n(n+1)}{2}$.

(j) E_{st}, $1 \leq s = t \leq n$, form a basis. Dimension is n.

(k) $\{I, A, A^2\}$ is a basis. Dimension is 3.

$H_n(\mathbb{C})$ and the set of normal matrices are not vector spaces over $\mathbb{C}$, since the former is not closed under the (complex) scalar multiplication, while the latter is not closed under the matrix addition. To see that $M_n(\mathbb{C}) = H_n(\mathbb{C}) + S_n(\mathbb{C})$, write $A = \frac{A+A^*}{2} + \frac{A-A^*}{2}$.

1.37 (a) All $n \times n$ matrices.

(b) All matrices of the form $\begin{pmatrix} a & b \\ 0 & a \end{pmatrix}$.

(c) All matrices of the form $\begin{pmatrix} c & 0 \\ 0 & d \end{pmatrix}$.

(d) All matrices of the form $\begin{pmatrix} a & b & c & d \\ 0 & a & b & c \\ 0 & 0 & a & b \\ 0 & 0 & 0 & a \end{pmatrix}$.

(e) All $n \times n$ scalar matrices cI_n.

1.38 First, we show that $S(A)$ is closed under the addition and scalar multiplication. Let $X, Y \in S(A)$ and c be a scalar. Then $A(X+Y) = AX + AY = 0$ and $A(cX) = c(AX) = 0$. So $S(A)$ is a subspace of $M_{n\times p}(\mathbb{C})$. When $m = n$ and if $X \in S(A^k)$, i.e., $A^kX = 0$, then $A^{k+1}X = A(A^kX) = 0$. Thus, $X \in S(A^{k+1})$. Hence, $S(A^k) \subseteq S(A^{k+1})$. Since each $S(A^k)$ is a subspace of $M_{n\times p}(\mathbb{C})$ and the dimension of $M_{n\times p}(\mathbb{C})$ is finite, there must exist a positive integer r such that $\dim S(A^r) = \dim S(A^{r+1})$. Hence, $S(A^r) = S(A^{r+1})$. We have

$$S(A) \subset S(A^2) \subset \cdots \subset S(A^r) = S(A^{r+1}) = \cdots$$

1.39 (a)$\Rightarrow$(b): Since $\operatorname{Im} A \subseteq \operatorname{Im} B$, every column vector of A is contained in $\operatorname{Im} B$, which is spanned by the columns of B. Thus every column of A is a linear combination of the columns of B; that is, (b).

(b)$\Rightarrow$(c): For a matrix X, denote the i-th column of X by X_i. Write

$$A = (A_1, A_2, \ldots, A_p), \quad B = (B_1, B_2, \ldots, B_q).$$

If (b) holds, then $A_i = c_{1i}B_1 + \cdots + c_{qi}B_q$ for some scalars c_{st}, where $s = 1, \ldots, q$, $t = 1, \ldots, p$. Taking $C = (c_{st})$ reveals $A = BC$.

(c)$\Rightarrow$(a): That $A = BC$ yields $A_i = BC_i = (B_1, \ldots, B_q)C_i$ for each $i = 1, 2, \ldots, p$; that is, each column A_i of A is a linear combination of the columns of B. Thus any linear combination of A_i is also a linear combination of the column vectors of B. Hence $\operatorname{Im} A \subseteq \operatorname{Im} B$.

1.40 (a) Easily check that $\operatorname{Ker} A$ is a subspace. Write $A = (a_1, a_2, \ldots, a_n)$, where all a_i are column vectors in $\mathbb{F}^n$. Then $\operatorname{Ker} A = \{0\}$ if and only if $Ax = 0$ has the unique solution $x = 0$; that is, $x_1a_1+x_2a_2+\cdots+x_na_n = 0$ if and only if $x_1 = x_2 = \cdots = x_n = 0$. If the columns are linearly independent, then $m \geq n$. In case $m > n$, then the rows of A are linearly dependent. In case $m = n$, then the rows of A are linearly independent.

(b) If $m < n$, then $r(A) \leq m < n$. So $\dim \operatorname{Ker} A = n - r(A) > 0$.

(c) If $Ax = 0$, then $A^2x = 0$.

(d) If $Ax = 0$, then $A^*Ax = 0$. So $\operatorname{Ker} A \subseteq \operatorname{Ker}(A^*A)$. Since $r(A^*A) = r(A)$, we have $\dim \operatorname{Ker} A = \dim \operatorname{Ker}(A^*A)$. It follows that $\operatorname{Ker} A = \operatorname{Ker}(A^*A)$.

(e) If $A = BC$ and $Ax = 0$, then $BCx = 0$. If B is invertible, then $Cx = B^{-1}BCx = 0$ and $\operatorname{Ker} A = \operatorname{Ker} C$.

1.41 We may assume $W_1 \neq W_2$. Take $\alpha_1 \in W_1$, $\alpha_1 \notin W_2$ and $\alpha_2 \notin W_1$, $\alpha_2 \in W_2$. Then $\alpha = \alpha_1 + \alpha_2 \notin W_1 \cup W_2$. To show that V has a basis that contains no vectors in W_1 and W_2, let $W_3 = \operatorname{Span}\{\alpha\}$. We claim that there exists a vector β that is not contained in any of W_1, W_2, W_3. To see this, pick $w_3 \notin W_3$ and consider $\beta_1 = \alpha + w_3$, $\beta_2 = \alpha + 2w_3$, and $\beta_3 = \alpha + 3w_3$. If they all fell in $W_1 \cup W_2$, then at least two would be in W_1 or W_2, say, β_1 and β_3 in W_2. It is immediate that $3\beta_1 - \beta_3 = 2\alpha \in W_2$ and $\alpha \in W_2$, a contraction. Let $\beta \notin W_1 \cup W_2 \cup W_3$. Then α and β are linearly independent. Now put $W_4 = \operatorname{Span}\{\alpha, \beta\}$. If $W_4 = V$, then we are done with the proof. Otherwise, pick $w_4 \notin W_4$ and consider $\beta + iw_4$, $i = 1, \ldots, 5$. In a similar way, there exists a vector $\gamma \in V$ that is not contained in any of the W's, and α, β, and γ are linearly independent. Inductively, there exists a basis of V such that no vector in the basis belongs to the subspaces W_1 and W_2.

In general, if $W_1, \ldots, W_m$ are nontrivial subspaces of a vector space V, there is a basis of V in which no vector falls in any of the subspaces.

1.42 Let W_1 and W_2 be subspaces of a finite dimensional vector space. If $\dim W_1 + \dim W_2 > \dim(W_1 + W_2)$, then, by the dimension identity, $W_1 \cap W_2 \neq \{0\}$. Note that $\dim W + \dim(\operatorname{Span}\{v_{i_1}, \ldots, v_{i_m}\}) = k+m > n$. There must be a nonzero vector in W and in the span of v_{i_j}'s.

1.43 (a) By definitions. The inclusions are nearly trivial.

(b) Take W_1 and W_2 to be the x-axis and the line $y = x$, respectively. Then $W_1 \cap W_2 = \{0\}$, while $W_1 + W_2$ is the entire xy-plane.

(c) In general, $W_1 \cup W_2$ is not a subspace; take the x- and y-axes. $W_1 \cup W_2$ is a subspace if and only if one of W_1 and W_2 is contained in the other: $W_1 \subseteq W_2$ or $W_2 \subseteq W_1$, i.e., $W_1 \cup W_2 = W_1 + W_2$.

(d) If S is a subspace containing W_1 and W_2, then every vector in the form $w_1 + w_2$, $w_1 \in W_1$, $w_2 \in W_2$, is contained in S. Thus $W_1 + W_2$ is contained in S.

1.44 (a) Let $u = (x_1, x_2, x_3, x_4)^t$, $v = (y_1, y_2, y_3, y_4)^t \in W$. Then, for any scalar λ, $\lambda u + v = (\lambda x_1 + y_1, \lambda x_2 + y_2, \lambda x_3 + y_3, \lambda x_4 + y_4)^t$, and $\lambda x_3 + y_3 = \lambda(x_1 + x_2) + (y_1 + y_2) = (\lambda x_1 + y_1) + (\lambda x_2 + y_2)$ and $\lambda x_4 + y_4 = \lambda(x_1 - x_2) + (y_1 - y_2) = \lambda(x_1 + y_1) - (\lambda x_2 + y_2)$. It follows that $\lambda u + v \in W$ and thus W is a subspace of $\mathbb{C}^4$.

(b) $(1, 0, 1, 1)^t$ and $(0, 1, 1, -1)^t$ form a basis of W. $\dim W = 2$.

(c) It is sufficient to notice that $(1, 0, 1, 1)^t \in W$;

1.45 Since $V_1 \cap V_2 \subseteq V_1 \subseteq V_1 + V_2$, $\dim(V_1 \cap V_2) \leq \dim(V_1) \leq \dim(V_1 + V_2)$. Thus the assumption $\dim(V_1 \cap V_2) + 1 = \dim(V_1 + V_2)$ implies that either $\dim(V_1) = \dim(V_1 \cap V_2)$ or $\dim(V_1) = \dim(V_1 + V_2)$. The former says $V_1 = V_1 \cap V_2$. Thus $V_1 \subseteq V_2$ and $V_2 = V_1 + V_2$. The latter ensures $V_1 = V_1 + V_2$. As a result, $V_2 \subseteq V_1$ and $V_2 = V_1 \cap V_2$.

1.46 For a counterexample, take W_1, W_2, and W_3 to be the x-, y-axes, and the line $y = x$, respectively. It does not contradict the set identity; the sum is usually "bigger" than the union. The former is a subspace, while the latter is not.

1.47 (a)$\Leftrightarrow$(b): If (a) holds, (b) is immediate. Assume (b). Let $w \in W_1 + W_2$ be written as $w = w_1 + w_2 = v_1 + v_2$, where w_1, $v_1 \in W_1$ and w_2, $v_2 \in W_2$. Then $(w_1 - v_1) + (w_2 - v_2) = 0$. By (b), $w_1 - v_1 = 0$, so $w_1 = v_1$. Likewise $w_2 = v_2$. This says the decomposition of w is unique. (b)$\Leftrightarrow$(c): If (b) holds and $w \in W_1 \cap W_2$, then $w + (-w) = 0$. By (b), $w = 0$. If (c) holds and $w_1 + w_2 = 0$, then $w_1 = -w_2 \in W_1 \cap W_2$. By (c), $w_1 = w_2 = 0$. (c)$\Leftrightarrow$(d): By the dimension identity.

For multiple subspaces $W_1, W_2, \dots, W_k$, $k \geq 3$, let $W = W_1 + W_2 + \cdots + W_k$. We say that W is a direct sum of W_1, W_2, $\dots$, W_k if for each $w \in W$, w can be expressed in exactly one way as a sum of vectors in $W_1, W_2, \dots, W_k$. The following statements are equivalent:

(i) W is a direct sum of $W_1, W_2, \dots, W_k$.

(ii) If $0 = w_1 + w_2 + \cdots + w_k$, $w_i \in W_i$, then all $w_i = 0$.

(iii) $\dim W = \dim W_1 + \dim W_2 + \cdots + \dim W_k$.

(iv) $W_i \cap \sum_{j \neq i} W_j = \{0\}$.

1.48 If $W_1 \neq V$, then $\dim(W_1) < \dim V$. Let $\{\alpha_1, \dots, \alpha_m, \alpha_{m+1}, \dots, \alpha_n\}$ be a basis of V, where $\alpha_1, \dots, \alpha_m \in W_1$. (This is possible since one may choose a basis for W_1 then extend it to a basis of V.) Set $W_2 = \text{Span}\{\alpha_{m+1}, \dots, \alpha_n\}$ and $W_3 = \text{Span}\{\alpha_1 + \alpha_{m+1}, \alpha_{m+2}, \dots, \alpha_n\}$. One may show that $V = W_1 \oplus W_2$ and $V = W_1 \oplus W_3$ with $W_2 \neq W_3$.

1.49 It is sufficient to notice that when w_i, $v_i \in W_i$, $i = 1, 2, 3$,

$$(w_1 + w_2 + w_3) + \lambda(v_1 + v_2 + v_3) = (w_1 + \lambda v_1) + (w_2 + \lambda v_2) + (w_3 + \lambda v_3)$$

again belongs to $W_1 + W_2 + W_3$.

For a counterexample, take the x-, y-axes, and the line $y = x$.

1.50 One may check that $W_1 \cap W_2 = \{0\}$ and

$$\begin{pmatrix} x & y \\ u & v \end{pmatrix} = \begin{pmatrix} \frac{x+v}{2} & \frac{y-u}{2} \\ -\frac{y-u}{2} & \frac{x+v}{2} \end{pmatrix} + \begin{pmatrix} \frac{x-v}{2} & \frac{y+u}{2} \\ \frac{y+u}{2} & -\frac{x-v}{2} \end{pmatrix}.$$

$V_2 = \{A \in M_n(\mathbb{R}) \mid A^t = -A\}$, the set of skew-symmetric matrices.

1.51 Let f and g be even functions. Then for any $r \in \mathbb{R}$,

$$(f + rg)(-x) = f(-x) + rg(-x) = f(x) + rg(x) = (f + rg)(x);$$

that is, $f + rg \in W_1$. So W_1 is a subspace. Similarly, W_2 is a subspace too. Now for any $f \in \mathcal{C}(\mathbb{R})$, we can write $f = f_e + f_o$, where

$$f_e = \frac{1}{2}\big(f(x) + f(-x)\big), \qquad f_o = \frac{1}{2}\big(f(x) - f(-x)\big).$$

Hence, $\mathcal{C}(\mathbb{R}) = W_1 + W_2$. Obviously, $W_1 \cap W_2 = \{0\}$. Thus $\mathcal{C}(\mathbb{R}) = W_1 \oplus W_2$. There are many functions that are neither even nor odd.

Hints and Answers for Chapter 2

2.1 $-48,\ -12(x+4),\ -x^{10}(1-x^2)(1-x^6)$.

2.2 6.

2.3 The zero block submatrix is too "big"; every expanded term is zero.

2.4 $(a_1b_2-a_2b_1)(a_3b_4-a_4b_3),\ (a_2a_3-b_2b_3)(a_1a_4-b_1b_4)$.

2.5 $a_1a_2a_3a_4a_5a_6$.

2.6 (a) Use induction on n. Subtracting the second column from the first and expanding the resulting determinant along the first column,

$$\Delta_n = (p_1-a)\Delta_{n-1} + a(p_2-b)\cdots(p_n-b).$$

By induction

$$\Delta_{n-1} = \frac{bF(a)-aF(b)}{b-a},$$

where $F(x) = (p_2-x)\cdots(p_n-x)$. Upon simplification,

$$\Delta_n = \frac{bf(a)-af(b)}{b-a}, \quad \text{if } a \neq b.$$

(b) If $a=b$, then

$$\begin{aligned}
\Delta_n &= (p_1-a)\Delta_{n-1} + af_1(a)\\
&= (p_1-a)[(p_2-a)\Delta_{n-2} + aF_2(a)] + af_1(a)\\
&= (p_1-a)(p_2-a)\Delta_{n-2} + af_2(a) + af_1(a)\\
&= \cdots\cdots\\
&= (p_1-a)\cdots(p_{n-2}-a)\Delta_2 + af_{n-2}(a) + \cdots + af_1(a).
\end{aligned}$$

Note that

$$\Delta_2 = p_np_{n-1} - a^2 = p_n(p_{n-1}-a) + (p_n-a)a.$$

The desired result follows immediately.

(c) $[a+(n-1)b](a-b)^{n-1}$.

2.7 Use induction on n. For the special case, take $a_i = i$.

2.8 Use induction on n. If $a = b$, the determinant is equal to $(n+1)a^n$.

2.9 $\lambda^{10} - 10^{10}$.

2.10 Expand $|\lambda I - A|$ along the first column.

2.11 Expand the first determinant, and then differentiate each term; $\operatorname{tr} A$.

2.12 Consider the first column of A. Multiply by -1 the rows with first entry -1, and then subtract the first row from other rows. The resulting matrix has entries only 0 and ± 2 except the first row.

2.13 $\det(A+B) = 40$; $\det C = 5$.

2.14 (a) $|A| = |A^t| = |-A| = (-1)^n|A| = -|A|$ if n is odd. So $|A| = 0$.

(b) $|A|^2 = |A^2| = |-I| = (-1)^n$. If n is odd, then $|A|^2 = -1$. This is impossible when A is a real matrix.

(c) No.

2.15 0, since $|A+I| = |A + AA^t| = |A|\,|I + A^t| = |A|\,|A+I|$.

2.16 Note that $XY = I$ implies $YX = I$ when X and Y are square.

2.17 $B = A^2 - 2A + 2I = A^2 - 2A + A^3 = A(A^2 + A - 2I) = A(A+2I)(A-I)$. However, $I = A^3 - I = (A-I)(A^2 + A + I)$. So $|A - I| \neq 0$. $A^3 + 8I = 10I$, also $(A+2I)(A^2 - 4A + 4I) = 10I$. So $|A + 2I| \neq 0$. It follows that $|B| = |A|\,|A+2I|\,|A-I| \neq 0$ and B is invertible.

2.18 $$\begin{pmatrix} 1 & 0 & 0 \\ 0 & 1 & 0 \\ -a & -b & 1 \end{pmatrix}, \quad \begin{pmatrix} -b/a & -c/d & -d/a & 1/a \\ 1 & 0 & 0 & 0 \\ 0 & 1 & 0 & 0 \\ 0 & 0 & 1 & 0 \end{pmatrix}.$$

2.19 $$\begin{pmatrix} 1 & -1 & 0 \\ 0 & 1 & -1 \\ 0 & 0 & 1 \end{pmatrix}, \quad \begin{pmatrix} 1 & -1 & 0 & \dots & 0 & 0 \\ 0 & 1 & -1 & \dots & 0 & 0 \\ \vdots & \ddots & \ddots & \ddots & \vdots & \vdots \\ \vdots & \vdots & \ddots & \ddots & \ddots & \vdots \\ 0 & 0 & \dots & 0 & 1 & -1 \\ 0 & 0 & \dots & 0 & 0 & 1 \end{pmatrix}.$$

2.20
$$\begin{pmatrix} 2 & -1 & 0 & \dots & 0 \\ -1 & 2 & -1 & \dots & 0 \\ 0 & -1 & 2 & \dots & 0 \\ \vdots & \vdots & \vdots & \ddots & \vdots \\ 0 & 0 & \dots & -1 & 2 \end{pmatrix}.$$

2.21 Determinant is $(-1)^{n-1}(n-1)$. The inverse is a matrix with diagonal entries $a = (2-n)/(n-1)$ and off-diagonal entries $b = 1/(n-1)$.

2.22
$$\begin{pmatrix} 0 & 0 & \dots & 0 & \frac{1}{a_n} \\ \frac{1}{a_1} & 0 & \dots & 0 & 0 \\ 0 & \frac{1}{a_2} & \dots & 0 & 0 \\ \vdots & \vdots & \ddots & \ddots & \vdots \\ 0 & 0 & \dots & \frac{1}{a_{n-1}} & 0 \end{pmatrix}.$$

2.23
$$\begin{pmatrix} A^{-1} & -A^{-1}BC^{-1} \\ 0 & C^{-1} \end{pmatrix}, \quad \begin{pmatrix} I & -X & XZ-Y \\ 0 & I & -Z \\ 0 & 0 & I \end{pmatrix}.$$

2.24 Use $V^{-1} = |V|^{-1}\operatorname{adj}(V)$ or apply elementary row operations to the augmented matrix (V, I) to get (I, V^{-1}).

$$V^{-1} = \frac{1}{|V|}\begin{pmatrix} a_3a_2(a_3-a_2) & -(a_3^2-a_2^2) & a_3-a_2 \\ -a_3a_1(a_3-a_1) & a_3^2-a_1^2 & -(a_3-a_1) \\ a_2a_1(a_2-a_1) & -(a_2^2-a_1^2) & a_2-a_1 \end{pmatrix}.$$

2.25 Check directly that $M^{-1}M = I$. (It would be harder to do MM^{-1}.)

2.26 Carefully verify that

$$(A-B)[A^{-1} + A^{-1}(B^{-1} - A^{-1})^{-1}A^{-1}] = I.$$

For the particular identity, substitute A by I and B by $-A$.

2.27 Multiply the right-hand side by $A + iB$. Then expand.

2.28 Since AB and CD are Hermitian, it is easy to verify that

$$\begin{pmatrix} A & B^* \\ C & D^* \end{pmatrix}\begin{pmatrix} D & -B \\ -C^* & A^* \end{pmatrix} = \begin{pmatrix} I & 0 \\ 0 & I \end{pmatrix}.$$

It follows that

$$\begin{pmatrix} D & -B \\ -C^* & A^* \end{pmatrix}\begin{pmatrix} A & B^* \\ C & D^* \end{pmatrix} = \begin{pmatrix} I & 0 \\ 0 & I \end{pmatrix},$$

and that, by observing the upper-left corner, $DA - BC = I$.

2.29 (a) If $A^*KA = K$, then A is nonsingular and $K = (A^{-1})^*KA^{-1}$. So $A^{-1} \in S_K$. By taking conjugate for both sides of $A^*KA = K$, we see $(\bar{A})^*K\bar{A} = K$. So $\bar{A} \in S_K$. From $(A^{-1})^*KA^{-1} = K$, taking inverses of both sides gives $AK^{-1}A^* = K^{-1}$. Note that $K^{-1} = K$. Thus $A^* \in S_K$. Consequently, $A^t = (\bar{A})^* \in S_K$.

(b) Since $(AB)^*K(AB) = B^*A^*KAB = B^*KB = K$, $AB \in S_K$. But kA and $A + B$ are not in S_K in general.

(c) (a) and (b) hold. In fact they hold for any K satisfying $K^2 = \pm I$.

2.30 (a) By the Laplace expansion theorem.

(b) 1, $(-1)^{mn}$, 1.

(c) Notice that

$$\begin{pmatrix} 0 & A \\ C & E \end{pmatrix} \begin{pmatrix} 0 & I_n \\ I_m & 0 \end{pmatrix} = \begin{pmatrix} A & 0 \\ E & C \end{pmatrix}.$$

By taking the determinant,

$$\begin{vmatrix} 0 & A \\ C & E \end{vmatrix} = (-1)^{mn}|A||C|.$$

2.31 Direct computation yields $S^2 = I$. So $S^{-1} = S$. It is obvious that $S^t = S$. $|S| = (-1)^{\frac{n^2+3n}{2}}$. The (i, j)-entry of SAS is $a_{n-i+1,n-j+1}$.

2.32 Notice that

$$\begin{pmatrix} 0 & I_n \\ I_m & 0 \end{pmatrix} \begin{pmatrix} A & B \\ C & D \end{pmatrix} \begin{pmatrix} 0 & I_p \\ I_q & 0 \end{pmatrix} = \begin{pmatrix} D & C \\ B & A \end{pmatrix}.$$

Taking the determinants of both sides gives the identity. When A, B, C, D are all square, say, $m \times m$, $mn + pq = 2m^2$ is an even number. So $(-1)^{(mn+pq)} = 1$. For the case of column and row vectors, $m = p$ and $n = q = 1$. Thus $mn + pq = 2m$ is also even. The identity holds. When B and C are switched, the two determinants may not equal.

2.33 It suffices to show $|A|\,|D| - |B|\,|C| = 0$. If A is invertible, then

$$\begin{pmatrix} I & 0 \\ -CA^{-1} & I \end{pmatrix} \begin{pmatrix} A & B \\ C & D \end{pmatrix} = \begin{pmatrix} A & B \\ 0 & D - CA^{-1}B \end{pmatrix}.$$

Since A is of rank n when A^{-1} exists,

$$D - CA^{-1}B = 0, \quad \text{or} \quad D = CA^{-1}B.$$

By taking determinant, $|A||D| - |B||C| = 0$.

If $|A| = 0$, it must be shown that $|B| = 0$ or $|C| = 0$. Suppose otherwise B (or C similarly) is invertible, then

$$\begin{pmatrix} I & 0 \\ -DB^{-1} & I \end{pmatrix} \begin{pmatrix} A & B \\ C & D \end{pmatrix} = \begin{pmatrix} A & B \\ C - DB^{-1}A & 0 \end{pmatrix}.$$

Since B is of rank n,

$$C - DB^{-1}A = 0, \quad \text{or} \quad C = DB^{-1}A$$

and

$$|C| = |DB^{-1}A| = |D||B^{-1}||A| = 0.$$

2.34 (a) Note that

$$\begin{pmatrix} A & B \\ C & D \end{pmatrix} \begin{pmatrix} D^t & 0 \\ -C^t & I \end{pmatrix} = \begin{pmatrix} AD^t - BC^t & B \\ CD^t - DC^t & D \end{pmatrix}.$$

Using $CD^t = DC^t$ and taking determinants, we have

$$\begin{vmatrix} A & B \\ C & D \end{vmatrix} |D^t| = |AD^t - BC^t|\,|D|.$$

If D is nonsingular, then the conclusion follows immediately by dividing both sides by $|D|$. Now suppose $|D| = 0$. If $C^t = C$,

$$C(D + \epsilon I)^t = (D + \epsilon I)C^t,$$

where $\epsilon > 0$. Using $D + \epsilon I$ for D in the above argument, we have

$$\begin{vmatrix} A & B \\ C & D + \epsilon I \end{vmatrix} = |A(D + \epsilon I)^t - BC^t|,$$

for all ϵ for which $D + \epsilon I$ is nonsingular. Notice that both sides of the above identity are continuous functions of ϵ and there are a finite number of ϵ for which $D + \epsilon I$ is singular. Letting $\epsilon \to 0$ yields the desired identity.

If C is not symmetric, let P and Q be the invertible such that

$$PCQ = \begin{pmatrix} I_r & 0 \\ 0 & 0 \end{pmatrix},$$

where r is the rank of C. Consider

$$\begin{pmatrix} I & 0 \\ 0 & P \end{pmatrix} \begin{pmatrix} A & B \\ C & D \end{pmatrix} \begin{pmatrix} Q & 0 \\ 0 & (Q^t)^{-1} \end{pmatrix}$$

$$= \begin{pmatrix} AQ & B(Q^t)^{-1} \\ PCQ & PD(Q^t)^{-1} \end{pmatrix}.$$

Note that PCQ is symmetric. Apply the earlier result.

(b) Take determinants of both sides of the matrix identity:

$$\begin{pmatrix} A & B \\ C & D \end{pmatrix} \begin{pmatrix} D^t & 0 \\ C^t & I \end{pmatrix} = \begin{pmatrix} AD^t + BC^t & B \\ CD^t + DC^t & D \end{pmatrix}.$$

(c) Take A, B, C, D to be, respectively,

$$\begin{pmatrix} 1 & 0 \\ 0 & 0 \end{pmatrix}, \begin{pmatrix} 0 & 0 \\ 0 & 1 \end{pmatrix}, \begin{pmatrix} 0 & 1 \\ 0 & 0 \end{pmatrix}, \begin{pmatrix} 0 & 0 \\ 1 & 0 \end{pmatrix}.$$

(d) $1^2 = (-1)^2$.

2.35 (a) If A^{-1} exists, then

$$\begin{pmatrix} I & 0 \\ -CA^{-1} & I \end{pmatrix} \begin{pmatrix} A & B \\ C & D \end{pmatrix} = \begin{pmatrix} A & B \\ 0 & D - CA^{-1}B \end{pmatrix}.$$

By taking determinant,

$$\begin{vmatrix} A & B \\ C & D \end{vmatrix} = |A|\,|D - CA^{-1}B|.$$

(b) Suppose $AC = CA$. If A^{-1} exists,

$$|A|\,|D - CA^{-1}B| = |AD - ACA^{-1}B| = |AD - CB|.$$

If A is not invertible, we take a positive number μ such that $|A + \varepsilon I| \neq 0$ for every ε, $0 < \varepsilon < \mu$.
Since $A + \varepsilon I$ and C commute,

$$\begin{vmatrix} A + \varepsilon I & B \\ C & D \end{vmatrix} = |(A + \varepsilon I)D - CB|.$$

Note that both sides are continuous functions of ε. Letting $\varepsilon \to 0$ results in the desired result for the case where A is singular.

(c) No. Take A, B, C, and D to be, respectively,

$$\begin{pmatrix} 1 & -1 \\ 0 & 0 \end{pmatrix}, \begin{pmatrix} 1 & 1 \\ 1 & 1 \end{pmatrix}, \begin{pmatrix} 1 & -1 \\ 0 & 0 \end{pmatrix}, \begin{pmatrix} 1 & 0 \\ 0 & 0 \end{pmatrix}.$$

Then $AC = CA$, $|AD - CB| = 0$, but $|AD - BC| = -1$.

(d) No. Take A, B, C, and D to be, respectively,

$$\begin{pmatrix} 1 & -1 \\ 0 & 0 \end{pmatrix}, \begin{pmatrix} 1 & 0 \\ -1 & 1 \end{pmatrix}, \begin{pmatrix} 1 & 1 \\ 1 & 1 \end{pmatrix}, \begin{pmatrix} 1 & 0 \\ 0 & 1 \end{pmatrix}.$$

Note that D commutes with other three matrices.

2.36 (a) No.

(b) It must be shown that there exist real numbers k_1, k_2, k_3, k_4, not all zero, such that $k_1B_1 + k_2B_2 + k_3B_3 + k_4B_4 = 0$. Let

$$A = \begin{pmatrix} a & b \\ c & d \end{pmatrix}, \quad B_i = \begin{pmatrix} w_i & x_i \\ y_i & z_i \end{pmatrix}, \quad i = 1, 2, 3, 4.$$

Then $|A + B_i| = |A| + |B_i|$ leads to

$$dw_i - cx_i - by_i + az_i = 0, \quad i = 1, 2, 3, 4.$$

Consider the linear equation of four unknowns w, x, y, z:

$$dw - cx - by + az = 0.$$

Since $A \neq 0$, say, $d \neq 0$, there are three free variables, and the solution space of the equation is of dimension 3. Thus any four vectors are linearly dependent; in particular, B_1, B_2, B_3, B_4 are linearly dependent.

2.37 For (a), $MM^{-1} = I$ implies $AY + BV = 0$ and $CY + DV = I$. Thus

$$\begin{pmatrix} A & B \\ C & D \end{pmatrix}\begin{pmatrix} I & 0 \\ 0 & V \end{pmatrix} = \begin{pmatrix} A & BV \\ C & DV \end{pmatrix} = \begin{pmatrix} A & -AY \\ C & I - CY \end{pmatrix}.$$

Taking determinant of both sides, we have (a):

$$\begin{aligned} |M|\,|V| &= \begin{vmatrix} A & -AY \\ C & I - CY \end{vmatrix} \\ &= \begin{vmatrix} A & 0 \\ C & I \end{vmatrix}\begin{vmatrix} I & -Y \\ 0 & I \end{vmatrix} \\ &= \begin{vmatrix} A & 0 \\ C & I \end{vmatrix} = |A|. \end{aligned}$$

To show (b), apply row-block operations to (M, I) to get (I, M^{-1}). The upper-left corner of this M^{-1} is $(D - CA^{-1}B)^{-1}$.

For (c), note that $U^{-1} = U^*$. And for (d), if M is real orthogonal, then $|A| = |V|$ or $|A| = -|V|$.

2.38 It is routine to verify (a), (b), and (c).

(d) $Z^n = r^n \begin{pmatrix} \cos n\theta & \sin n\theta \\ -\sin n\theta & \cos n\theta \end{pmatrix}$, since $z^n = r^n(\cos n\theta + i \sin n\theta)$.

(e) $\begin{pmatrix} 0 & 1 \\ -1 & 0 \end{pmatrix}$.

(f) Check $ZZ^{-1} = I$.

(g) Direct verification.

(h) $|Q| = |u|^2 + |v|^2 \geq 0$. $Q^{-1} = \begin{pmatrix} \bar{u} & -v \\ \bar{v} & u \end{pmatrix}$ when $|Q| = 1$.

(i) Write $u = u_1 + iu_2$ and $v = v_1 + iv_2$ to get R. Note that U and $-V^t$ commute. By computation,

$$|R| = |UU^t + V^tV| = (|u|^2 + |v|^2)^2 \geq 0.$$

(j) Exchange the last two rows and columns of R.

(k) From (i), $|R| = 0 \Leftrightarrow |Q| = 0 \Leftrightarrow u = v = 0$.

2.39 Notice that

$$\begin{pmatrix} I & iI \\ 0 & I \end{pmatrix} \begin{pmatrix} A & B \\ -B & A \end{pmatrix} \begin{pmatrix} I & -iI \\ 0 & I \end{pmatrix} = \begin{pmatrix} A - iB & 0 \\ -B & A + Bi \end{pmatrix}.$$

Thus

$$\begin{vmatrix} A & B \\ -B & A \end{vmatrix} = \begin{vmatrix} A - iB & 0 \\ -B & A + iB \end{vmatrix} = \overline{|A + iB|}|A + iB| \geq 0.$$

If A and B are complex matrices, then it is expected that

$$\begin{vmatrix} A & B \\ -\overline{B} & \overline{A} \end{vmatrix} \geq 0.$$

[Note: To prove this, a more advanced result that $\overline{A}A$ is similar to R^2 for some real matrix R is needed.]

2.40 Add the second column (matrices) to the first column, and then subtract the first row (matrices) from the second row:

$$\begin{vmatrix} A & B \\ B & A \end{vmatrix} = \begin{vmatrix} A + B & B \\ B + A & A \end{vmatrix} = \begin{vmatrix} A + B & B \\ 0 & A - B \end{vmatrix} = |A + B|\,|A - B|.$$

To find the inverse matrix, let $C = A + B$, $D = A - B$, and consider

$$\begin{pmatrix} A & B \\ B & A \end{pmatrix}\begin{pmatrix} X & Y \\ U & V \end{pmatrix} = \begin{pmatrix} I & 0 \\ 0 & I \end{pmatrix}.$$

Multiply this out:

$$AX + BU = I, \quad AY + BV = 0, \quad BX + AU = 0, \quad BY + AV = I.$$

Adding the first equation to the third reveals $(A + B)(X + U) = I$. So $X + U = C^{-1}$. Subtracting the first equation from the third gives $(A - B)(X - U) = I$. Thus $X - U = D^{-1}$. It follows that

$$X = \frac{1}{2}(C + D), \qquad U = \frac{1}{2}(C - D).$$

In a similar way, one can show that $Y = U$ and $V = X$. Thus

$$\begin{pmatrix} A & B \\ B & A \end{pmatrix}^{-1} = \frac{1}{2}\begin{pmatrix} C^{-1} + D^{-1} & C^{-1} - D^{-1} \\ C^{-1} - D^{-1} & C^{-1} + D^{-1} \end{pmatrix}.$$

2.41 (a) follows by observing

$$\begin{vmatrix} I & x \\ y^* & 1 \end{vmatrix} = \begin{vmatrix} I & x \\ 0 & 1 - y^*x \end{vmatrix} = 1 - y^*x$$

and

$$\begin{vmatrix} I & x \\ y^* & 1 \end{vmatrix} = \begin{vmatrix} I - xy^* & 0 \\ y^* & 1 \end{vmatrix} = |I - xy^*|.$$

For (b), it is sufficient to notice that

$$\begin{pmatrix} 0 & 1 \\ I & 0 \end{pmatrix}\begin{pmatrix} I & x \\ y^* & 1 \end{pmatrix}\begin{pmatrix} 0 & I \\ 1 & 0 \end{pmatrix} = \begin{pmatrix} 1 & y^* \\ x & I \end{pmatrix}.$$

For (c) and (d), one may verify directly through multiplications.

2.42 If $r(A) = 1$, then any two rows of A are linearly dependent. Some row of A is not zero, say, the first row. Then all other rows are multiples of the first row. The conclusion follows immediately.

2.43 If $r(A) = n$, then A is invertible. Let $x_1(Au_1) + x_2(Au_2) + \cdots + x_n(Au_n) = 0$. Premultiplying both sides by A^{-1} shows $x_1u_1 + x_2u_2 + \cdots + x_nu_n = 0$. Thus $x_1 = x_2 = \cdots = x_n = 0$ since $u_1, u_2, \ldots, u_n$ are linearly independent. So $Au_1, Au_2, \ldots, Au_n$ are linearly independent. For the other direction, if $Au_1, Au_2, \ldots, Au_n$ are linearly dependent, let $y_1(Au_1)+y_2(Au_2)+\cdots+y_n(Au_n)=0$, where not all y are zero. Then $A(y_1u_1+y_2u_2+\cdots+y_nu_n)=0$. Since $u_1,u_2,\dots,u_n$ are linearly independent, $y_1u_1+y_2u_2+\cdots+y_nu_n\neq 0$. Thus the system $Ax=0$ has a nonzero solution, and A is singular.

2.44 Use row and column operations on A. Each elementary operation results in an elementary (real if A is real) matrix that is invertible.

2.45 (a) False.

(b) False.

(c) True for $k\geq 1$. False otherwise.

(d) False.

(e) False.

(f) False.

2.46 (a) $(1,i)=i(-i,1)$.

(b) Let $x(1,i)+y(-i,1)=0$, where x and y are real. Then $x-yi=0$. Thus $x=y=0$, and $(1,i)$ and $(-i,1)$ are linearly independent.

(c) No. In fact, A is not invertible.

(d) $U^*AU=\operatorname{diag}(2,0)$.

(e) $r(A)=1$.

(f) Since $x\in\mathbb{R}^2\subseteq\mathbb{C}^2$, $Ax\in W_{\mathbb{C}}$ for all $x\in\mathbb{R}^2$.

(g) $W_{\mathbb{C}}$ is $\operatorname{Im} A$. It is a subspace of $\mathbb{C}^2$ over $\mathbb{C}$, thus also over $\mathbb{R}$.

(h) $a(Ax)+b(Ay)=A(ax+by)$. When a,b,x,y are real, $ax+by$ is real. Thus $W_{\mathbb{R}}$ is a subspace of $\mathbb{C}^2$ over $\mathbb{R}$. When a,b are complex, $ax+by$ may not be real. So $W_{\mathbb{R}}$ is not closed over $\mathbb{C}$.

(i) $\dim W_{\mathbb{R}}=2$ over $\mathbb{R}$:

$$A(x,y)^t=(x+iy,-ix+y)^t=x(1,-i)^t+y(i,1)^t.$$

Similarly, $\dim W_{\mathbb{C}}=2$ over $\mathbb{R}$, because $A(x,y)^t=(a-d)(1,-i)^t+(b+c)(i,1)^t$, where $x=a+bi$, $y=c+di$, a, b, c, d are real, and $\dim W_{\mathbb{C}}=1$ over $\mathbb{C}$, for $A(x,y)^t=c(1,-i)^t$, $c=x+iy$.

2.47 (a) $A^*=B^*-iC^*=B^t-iC^t=B+iC$. So $B^t=B$ and $C^t=-C$.

(b) $x^tAx=x^tBx+ix^tCx$. Since A is Hermitian, x^tAx is always real for $x\in\mathbb{R}^n$, as is x^tBx. Therefore, $x^tAx=x^tBx$ and $ix^tCx=0$.

(c) $Ax = Bx+iCx$. If $Ax = 0$ and x is real, then $0 = Ax = Bx+iCx$. Thus $Bx = Cx = 0$.

(d) Take $x = (1, i)^t$. Then $Ax = 0$, so $x^*Ax = 0$; $x^*Bx = x^*x > 0$.

(e) Take $x = (1, -1)^t$. $r(A) = 2$, $r(B) = 1$.

2.48 This can be seen by taking $A = \begin{pmatrix} I_r & 0 \\ 0 & 0 \end{pmatrix}$ (see Problem 2.44).

2.49 3.

2.50 -3.

2.51 Since $AB = 0$, the column vectors of B are contained in

$$\operatorname{Ker} A = \{\, x \mid Ax = 0 \,\}.$$

Since

$$r(A) + \dim \operatorname{Ker} A = n,$$

where n is the number of unknowns, it follows that

$$r(A) + r(B) \le r(A) + \dim \operatorname{Ker} A = n.$$

2.52 Let A_s be the submatrix of A by deleting s rows from A. Then

$$r(A) - s \le r(A_s).$$

Similarly,

$$r(A_s) - t \le r(B).$$

Thus

$$r(A) \le s + t + r(B).$$

2.53 Each column of AB is a linear combination of the columns of A. So $r(AB) \le r(A)$. Considering rows gives $r(AB) \le r(B)$. Thus

$$r(AB) \le \min\{r(A), r(B)\}.$$

The columns of $A + B$ are linear combinations of those of A, B. So

$$r(A + B) \le r(A) + r(B).$$

We now show a general rank inequality: $r(A) + r(B) \le r(AB) + n$, where A is $p \times n$ and B is $n \times q$. Notice that

$$\begin{pmatrix} I_n & 0 \\ -A & I_p \end{pmatrix} \begin{pmatrix} I_n & B \\ A & 0 \end{pmatrix} \begin{pmatrix} I_n & -B \\ 0 & I_p \end{pmatrix} = \begin{pmatrix} I_n & 0 \\ 0 & -AB \end{pmatrix}.$$

Thus

$$r\begin{pmatrix} I_n & B \\ A & 0 \end{pmatrix} = r\begin{pmatrix} I_n & 0 \\ 0 & -AB \end{pmatrix} = n + r(AB).$$

The desired rank inequality follows immediately by noting that

$$r(A) + r(B) \leq r\begin{pmatrix} I_n & B \\ A & 0 \end{pmatrix}.$$

2.54 We use the rank identity $r(XY) \geq r(X) + r(Y) - n$, where X is $p \times n$ and Y is $n \times q$, to show the rank identity for three matrices.

Let $r(B) = r$. Then there exist invertible matrices P and Q such that

$$B = P\begin{pmatrix} I_r & 0 \\ 0 & 0 \end{pmatrix} Q = MN, \quad \text{where } P = (M, S), \ \ Q = \begin{pmatrix} N \\ T \end{pmatrix},$$

where M is $n \times r$ and N is $r \times q$. By Problem 2.53, we have

$$\begin{aligned} r(ABC) &= r(AMNC) \geq r(AM) + r(NC) - r \\ &\geq r(AMN) + r(MNC) - r \\ &= r(AB) + r(BC) - r(B). \end{aligned}$$

2.55 Let W_1, W_2, W_3, and W_4 be the column spaces of A, B, $A + B$, and AB, respectively. Since $W_3 \subseteq W_1 + W_2$, we have

$$\dim W_3 \leq \dim(W_1 + W_2) = \dim W_1 + \dim W_2 - \dim(W_1 \cap W_2)$$

or

$$r(A + B) \leq r(A) + r(B) - \dim(W_1 \cap W_2).$$

We claim that $r(AB) = \dim W_4 \leq \dim(W_1 \cap W_2)$. To do so, we show $W_4 \subseteq W_1 \cap W_2$. Write $B = (b_1, b_2, \ldots, b_n)$. Then $AB = (Ab_1, Ab_2, \ldots, Ab_n)$. Since each $Ab_i \in W_1$, we see $W_4 \subseteq W_1$. Given that $AB = BA$, we have similarly $W_4 \subseteq W_2$. Thus $W_4 \subseteq W_1 \cap W_2$.

2.56 By the Problem 2.53, we have

$$\begin{aligned} 0 &= r(A_1 A_2 \cdots A_k) \\ &\geq r(A_1) + r(A_2 \cdots A_k) - n \\ &\geq r(A_1) + r(A_2) + r(A_3 \cdots A_k) - 2n \\ &\geq \cdots \\ &\geq r(A_1) + r(A_2) + \cdots + r(A_k) - (k-1)n \end{aligned}$$

2.57 Consider the column spaces of (X, Z), (Z, Y), and Z. Denote them by W_1, W_2, and W_3, respectively. By dimension identity,

$$\begin{aligned}\dim(W_1 + W_2) &= \dim W_1 + \dim W_2 - \dim(W_1 \cap W_2) \\ &= r(X, Z) + r(Z, Y) - \dim(W_1 \cap W_2).\end{aligned}$$

Note that the column space of (X, Y) is contained in $W_1 + W_2$ and that $W_3 \subseteq W_1 \cap W_2$. So $r(X, Y) \leq \dim(W_1 + W_2)$ and $r(Z) \leq \dim(W_1 \cap W_2)$. The desired inequality follows.

2.58 It is sufficient to notice that $(A^*A)x = 0$ and $Ax = 0$ have the same solution space since $Ax = 0 \Leftrightarrow x^*A^*Ax = 0$.

$r(A^tA)$ is not equal to $r(A)$ in general. Take $A = \begin{pmatrix} 1 & i \\ i & -1 \end{pmatrix}$. Then $r(A^tA) = 0$, but $r(A) = 1$. Similarly, $r(\bar{A}A) \neq r(A)$.

2.59 M_2.

2.60 There are two important facts regarding $\operatorname{adj}(A)$:

$$A \operatorname{adj}(A) = \operatorname{adj}(A)A = |A|I$$

and when A is invertible,

$$\operatorname{adj}(A) = |A|A^{-1}.$$

(a) $\operatorname{adj}(A)$ is invertible if and only if $|A| \neq 0$.

(b) $A \operatorname{adj}(A) = 0$ implies that the column vectors of $\operatorname{adj}(A)$ are the solutions of $Ax = 0$. If $r(A) = n-1$, then $\dim \operatorname{Ker} A = 1$ and the column vectors of $\operatorname{adj}(A)$ are mutually linearly dependent. Thus $r(\operatorname{adj}(A)) = 1$. The other direction is a part of (c).

(c) Consider $(n-1) \times (n-1)$ minors of A.

(d) It follows from the two facts given above.

(e) By the first fact mentioned above

$$\operatorname{adj}(A) \cdot \operatorname{adj}(\operatorname{adj}(A)) = |\operatorname{adj}(A)|I = |A|^{n-1}I.$$

Replace the left-most $\operatorname{adj}(A)$ by $|A|A^{-1}$ when A is invertible. If A is singular, then both sides vanish.

(f) First consider the case where A and B are nonsingular. For the singular case, use $A + \varepsilon I$ and $B + \varepsilon I$ to substitute A and B, respectively, then apply an argument of continuity.

(g) By (f).

(h) Use definition.

(i) $A_{ij} = A^*_{ji}$ when A is Hermitian.

$$\overbrace{\mathrm{adj}\cdots\mathrm{adj}}^{k}(A) = A \text{ when } k \text{ is even, } A^{-1} \text{ when } k \text{ is odd.}$$

If A is invertible and has eigenvalues $\lambda_1, \lambda_2, \ldots, \lambda_n$, then the eigenvalues of $\mathrm{adj}(A)$ are

$$\frac{1}{\lambda_1}|A|, \ \frac{1}{\lambda_2}|A|, \ \ldots, \ \frac{1}{\lambda_n}|A|.$$

If A is singular, the eigenvalues of $\mathrm{adj}(A)$ are

$$0, \ 0, \ \ldots, \ 0, \quad \text{and} \quad \sum_{i=1}^{n} |A_{ii}|.$$

2.61 It suffices to show that $Ax = 0$ has only the trivial solution 0. Suppose that $Ax = 0$ has a nonzero solution $x = (k_1, k_2, \ldots, k_n)$. Let

$$|k_s| = \max_{1\le i\le n}\{|k_i|\}.$$

Then $|k_s| \neq 0$. However, the s-th equation of $Ax = 0$ is

$$a_{s1}k_1 + a_{s2}k_2 + \cdots + a_{ss}k_s + \cdots + a_{sn}k_n = 0.$$

Thus

$$a_{ss}k_s = -\sum_{j=1,\ j\neq s}^{n} a_{sj}k_j$$

and

$$|a_{ss}| \le \sum_{j=1,\ j\neq s}^{n} \left|a_{sj}\frac{k_j}{k_s}\right| \le \sum_{j=1,\ j\neq s}^{n} |a_{sj}|,$$

a contradiction to the given condition.

2.62 Use Problem 2.54 in the first inequality below:

$$\begin{aligned} r(A-AB)+r(B-AB) &= r(A^2-AB)+r(B^2-AB) \\ &= r(A(A-B))+r((B-A)B) \\ &= r(A(A-B))+r((A-B)B) \\ &\leq r(A(A-B)B)+r(A-B) \\ &= r(A^2B-AB^2)+r(A-B) \\ &= r(AB-AB)+r(A-B) \\ &= r(A-B) \\ &\leq r(A-AB)+r(AB-B) \\ &= r(A-AB)+r(B-BA). \end{aligned}$$

2.63 $I = A+(I-A) \Rightarrow n = r(A+(I-A)) \leq r(A)+r(I-A)$ and $A^2 = A \Leftrightarrow A(A-I) = 0 \Leftrightarrow \operatorname{Im}(A-I) \subseteq \operatorname{Ker} A \Rightarrow r(A-I) \leq n-r(A)$.

2.64 It is sufficient to notice that

$$\begin{pmatrix} I_m & -A \\ 0 & I_n \end{pmatrix}\begin{pmatrix} 0 & A \\ B & I_n \end{pmatrix}\begin{pmatrix} I_p & 0 \\ -B & I_n \end{pmatrix} = \begin{pmatrix} -AB & 0 \\ 0 & I_n \end{pmatrix}.$$

2.65 Notice that

$$\begin{pmatrix} I_m & 0 \\ -A^* & I_n \end{pmatrix}\begin{pmatrix} I_m & A \\ A^* & I_n \end{pmatrix}\begin{pmatrix} I_m & -A \\ 0 & I_n \end{pmatrix} = \begin{pmatrix} I_m & 0 \\ 0 & I_n - A^*A \end{pmatrix}.$$

So

$$r\begin{pmatrix} I_m & A \\ A^* & I_n \end{pmatrix} = m + r(I_n - A^*A).$$

Similarly,

$$r\begin{pmatrix} I_m & A \\ A^* & I_n \end{pmatrix} = n + r(I_m - AA^*).$$

Thus

$$r(I_m - AA^*) - r(I_n - A^*A) = m - n.$$

2.66 (a) $\lambda \neq -3$. (b) $\lambda = -3$.

2.67 There are three possibilities for $A^* = \operatorname{adj}(A)$:

(1) $A = 0$.

(2) A is a unitary matrix and $|A| = 1$.

(3) A is a 2×2 matrix of the form $\begin{pmatrix} a & b \\ -\bar{b} & \bar{a} \end{pmatrix}$.

2.68 $\lambda \neq 1$.

2.69 $\lambda = 1$.

2.70 $\alpha = (-1, 1, 0, 0, 0)$ and $\beta = (-1, 0, -1, 0, 1)$ form a basis for the solution space. Thus the general solution is $x = \lambda\alpha + \mu\beta$, λ, $\mu \in \mathbb{F}$.

2.71 The dimension is 2. $\eta_1 = (-\frac{3}{2}, \frac{7}{2}, 1, 0)$ and $\eta_2 = (-1, -2, 0, 1)$ form a basis for the solution space. The general solution is $\eta = x_1\eta_1 + x_2\eta_2$, where x_1 and x_2 are scalars.

2.72 Adding all equations gives $(x_1 + x_2 + x_3 + x_4 + x_5)(y - 2) = 0$.

If $y = 2$, $x_1 = x_2 = x_3 = x_4 = x_5 = c$, where c is any number.

If $y \neq 2$, by eliminating x_5, x_4, and x_3 one by one from the given equations, one has $(y^2 + y - 1)(x_2 - x_1) = 0$ and

$$(y^2 + y - 1)[x_2 - (y - 1)x_1] = 0.$$

Thus if $y \neq 2$ and $y^2 + y - 1 \neq 0$, then $x_1 = x_2 = x_3 = x_4 = x_5 = 0$.

If $y \neq 2$ and $y^2 + y - 1 = 0$, the solution is

$$\begin{aligned} x_1 &= s \\ x_2 &= t \\ x_3 &= yt - s \\ x_4 &= (y^2 - 1)t - ys \\ x_5 &= ys - t, \end{aligned}$$

where s, t are arbitrary, y is a solution to $y^2 + y - 1 = 0$.

An alternative approach is to apply elementary row operations to the coefficient matrix

$$\begin{pmatrix} -y & 1 & 0 & 0 & 1 \\ 1 & -y & 1 & 0 & 0 \\ 0 & 1 & -y & 1 & 0 \\ 0 & 0 & 1 & -y & 1 \\ 1 & 0 & 0 & 1 & -y \end{pmatrix}.$$

2.73 Apply elementary row operations to the coefficient matrix to get

$$\begin{pmatrix} a & 1 & 1 & 1 \\ 0 & b-1 & 1 & 0 \\ 0 & 0 & b+1 & 2(b-1) \end{pmatrix}.$$

There are six cases:

$b = 1$: infinite solution.

$b = 5$, $a = 0$: infinite solution.

$b = 5$, $a \neq 0$: unique solution.

$b = -1$: no solution.

$b \neq \pm 1, 5$, $a \neq 0$: unique solution.

$b \neq 1, 5$, $a = 0$: no solution.

2.74 Write the complex solution as $x = x_1 + ix_2$. Then $Ax_1 = 0$ and $Ax_2 = 0$. Either x_1 or x_2 is nonzero since x is nonzero.

2.75 Use elementary row operations to the coefficient matrix.

The dimension of the solution space is $n-1$ and the following vectors in $\mathbb{R}^{2n}$ form a basis for the solution space:

$$\begin{aligned} \epsilon_1 &= (-1, 1, 0, \ldots, 0, -1, 1, 0, \ldots, 0) \\ \epsilon_2 &= (-1, 0, 1, \ldots, 0, -1, 0, 1, \ldots, 0) \\ &\vdots \\ \epsilon_{n-1} &= (-1, 0, \ldots, 1, -1, 0, \ldots, 0, 1). \end{aligned}$$

2.76 (a) It is sufficient to notice that $r(A)$ is equal to the largest number of column vectors, which are linearly independent.

(b) Since $r(X) = r(PX)$ for any matrix X when P is invertible, it follows that $r[(\alpha_{i_1}, \alpha_{i_2}, \ldots, \alpha_{i_r})] = r[(\beta_{i_1}, \beta_{i_2}, \ldots, \beta_{i_r})]$.

(c) Apply elementary row operations to $(\gamma_1, \gamma_2, \gamma_3, \gamma_4)$ to get

$$(\gamma_1, \gamma_2, \gamma_3, \gamma_4) \longrightarrow \begin{pmatrix} 1 & 1 & 1 & 1 \\ 0 & 1 & 1 & 0 \\ 0 & 0 & 1 & 1 \\ 0 & 0 & 0 & 0 \end{pmatrix}.$$

Thus the dimension is 3. $\{\gamma_1, \gamma_2, \gamma_3\}$ is a basis. In fact, any three of $\gamma_1, \gamma_2, \gamma_3, \gamma_4$ form a basis.

2.77 For $W_1 \cap W_2$, consider the equation system

$$x_1\alpha_1 + x_2\alpha_2 + x_3\alpha_3 = y_1\beta_1 + y_2\beta_2.$$

The dimensions of $W_1, W_2, W_1 \cap W_2$ and $W_1 + W_2$ are 3, 2, 1, 4, respectively. $\{\beta_1\}$ is a basis for $W_1 \cap W_2$ and $\{\alpha_1, \alpha_2, \alpha_3, \beta_2\}$ is a basis for $W_1 + W_2$, which is spanned by α_1, α_2, α_3 and β_1, β_2.

2.78 Write the equation system as $Ax = b$. Take $b = e_i$, where e_i is the column vector with i-th component 1 and everywhere else 0, $i = 1, 2, \ldots, n$. Then there are column vectors C_i with integer components such that $AC_i = e_i$ for each i. Thus $AC = I$, where $C = (C_1, C_2, \ldots, C_n)$. Taking determinants, $|AC| = |A||C| = 1$, so $|A| = \pm 1$.

2.79 $\dim W_1 = n - 1, \dim W_2 = 1$, and $W_1 \cap W_2 = \{0\}$.

2.80 If $|A| = 0$, then $Ax = 0$ has a nonzero solution x_0. Let $B = (x_0, 0) \in M_n(\mathbb{C})$. Then $AB = 0$. If $AB = 0$ for some nonzero matrix B, then $Ab = 0$ for any column vector b of B. Thus $Ax = 0$ has a nonzero solution and A is singular; that is, $|A| = 0$.

2.81 It suffices to show $\operatorname{Ker} A \cap \operatorname{Ker} B \neq \{0\}$. By the dimension identity,

$$\begin{aligned}
&\dim(\operatorname{Ker} A \cap \operatorname{Ker} B) \\
&\quad = \dim \operatorname{Ker} A + \dim \operatorname{Ker} B - \dim(\operatorname{Ker} A + \operatorname{Ker} B) \\
&\quad = n - r(A) + n - r(B) - \dim(\operatorname{Ker} A + \operatorname{Ker} B) \\
&\quad = (n - r(A) - r(B)) + (n - \dim(\operatorname{Ker} A + \operatorname{Ker} B) \\
&\quad > n - \dim(\operatorname{Ker} A + \operatorname{Ker} B) \geq 0.
\end{aligned}$$

2.82 $r(A) = n - l$, $r(B) = n - m$. So $r(AB) \leq \min\{n - l, n - m\}$. Thus $\dim \operatorname{Ker}(AB) = n - r(AB) \geq \max\{n - (n - l), n - (n - m)\} = \max\{l, m\}$. If all $x \in \mathbb{F}^n$ fall in either $\operatorname{Ker} A$ or $\operatorname{Ker} B$; that is, $\mathbb{F}^n = \operatorname{Ker} A \cup \operatorname{Ker} B$, then $\operatorname{Ker} A = \mathbb{F}^n$ or $\operatorname{Ker} B = \mathbb{F}^n$, so $A = 0$ or $B = 0$.

2.83 First notice that $\operatorname{Ker} A \subseteq \operatorname{Ker}(A^2)$. If $r(A) = r(A^2)$, then $\dim \operatorname{Ker} A = \dim \operatorname{Ker}(A^2)$. This implies $\operatorname{Ker} A = \operatorname{Ker}(A^2)$.

2.84 It is sufficient to show that there is an invertible matrix C such that $A = CB$ when $Ax = 0$ and $Bx = 0$ have the same solution space.

First notice that A and B must have the same rank, denoted by r. Let P_1 and P_2 be the permutation matrices such that

$$P_1 A = \begin{pmatrix} A_1 \\ Q_1 A_1 \end{pmatrix} \quad \text{and} \quad P_2 B = \begin{pmatrix} B_1 \\ Q_2 B_1 \end{pmatrix},$$

where A_1 and B_1 are, respectively, $r \times n$ submatrices of A and B with rank r, and Q_1 and Q_2 are some matrices of size $(m - r) \times n$. The systems $Ax = 0$ and $Bx = 0$ have the same solution space if and only if $A_1x = 0$ and $B_1x = 0$ have the same solution space. Since

$$r\begin{pmatrix} A_1 \\ B_1 \end{pmatrix} = r(B_1),$$

there is an $r \times r$ invertible matrix C_1 such that $A_1 = C_1B_1$. Thus

$$\begin{aligned} P_1A &= \begin{pmatrix} A_1 \\ Q_1A_1 \end{pmatrix} \\ &= \begin{pmatrix} C_1 & 0 \\ Q_1C_1 - Q_2 & I_{m-r} \end{pmatrix}\begin{pmatrix} B_1 \\ Q_2B_1 \end{pmatrix} \\ &= C_2P_2B, \end{aligned}$$

where

$$C_2 = \begin{pmatrix} C_1 & 0 \\ Q_1C_1 - Q_2 & I_{m-r} \end{pmatrix}$$

is of full rank. Take $C = P_1^{-1}C_2P_2$. Then $A = CB$.

If $r(A^2) = r(A)$, then $A^2x = 0$ and $Ax = 0$ have the same solution space. From the above result, $A^2 = DA$ for some invertible D.

2.85 Suppose that $\lambda_1\eta_1 + \cdots + \lambda_n\eta_n$ is a solution. Then

$$A(\lambda_1\eta_1) + \cdots + A(\lambda_n\eta_n) = b.$$

Since $A(\lambda_i\eta_i) = \lambda_i b$, $i = 1, \dots, n$, we have $(\lambda_1 + \cdots + \lambda_n)b = b$. Thus, $\lambda_1 + \cdots + \lambda_n = 1$ for $b \neq 0$. Conversely, if $\lambda_1 + \cdots + \lambda_n = 1$, then

$$A(\lambda_1\eta_1 + \cdots + \lambda_n\eta_n) = \lambda_1A\eta_1 + \cdots + \lambda_nA\eta_n = (\lambda_1 + \cdots + \lambda_n)b = b.$$

2.86 $r(A^*A) \leq r(A^*A, A^*b) = r((A^*(A, b)) \leq r(A^*)$. However, $r(A^*A) = r(A^*)$. So $r(A^*A) = r(A^*A, A^*b)$. Thus the coefficient matrix A^*A and the augmented matrix (A^*A, A^*b) have the same rank. It follows that $A^*Ax = A^*b$ is consistent.

2.87 $|\tilde{A}| = 0$. So (d) is right.

2.88 Let $A = (a_{ij})$, $c = (c_1, c_2, \dots, c_n)$, $b = (b_1, b_2, \dots, b_n)^t$. Then the augmented matrices of the equation systems are, respectively,

$$M = \begin{pmatrix} A & b \\ c & d \end{pmatrix}, \quad M^t = \begin{pmatrix} A^t & c^t \\ b^t & d \end{pmatrix}.$$

Since A is nonsingular, $r(A) = n$, we see $r(M) = n$ or $n+1$. The first equation system has a solution if and only if $r(M) = n$. The second system has a solution if and only if $r(M^t) = n$. However, $r(M) = r(M^t)$. The two systems will both have solution or no solution. In case they both have solution, the solution to the first system is $x = A^{-1}b$ and the solution to the second system is $y = (A^t)^{-1}c^t$.

2.89 No. Let A be an $n \times n$ matrix. Since $Ax = 0$ has nonzero solutions, $r(A) < n$. Note that $r(A) = r(A^t)$. Let $B = (A^t, b)$. If $r(B) \neq r(A)$, then $A^t x = b$ has no solution. If $r(B) = r(A^t) = r(A) < n$, then there are infinitely many solutions. In either case, $A^t x = b$ cannot have a unique solution.

2.90 If B is nonsingular, then it is obvious. Let the rank of B be r and first consider the case $B = \begin{pmatrix} I_r & 0 \\ 0 & 0 \end{pmatrix}$. Since $r(AB) = r(A)$, the first r columns of A spans the column space of A. So we may write A as $A = (A_r, A_r C)$. $X_1 AB = X_2 AB$ implies $X_1 A_r = X_2 A_r$. Thus $X_1 A = (X_1 A_r, X_1 A_r C) = (X_2 A_r, X_2 A_r C) = X_2 A$. For a general B, let $B = P \begin{pmatrix} I_r & 0 \\ 0 & 0 \end{pmatrix} Q$, P, Q invertible, and apply the above argument.

2.91 Assume that the three lines are different from each other. Let the lines intersect at (x_0, y_0). Adding the three equations, we have

$$(x_0 + y_0 + 1)(a + b + c) = 0.$$

We show that $x_0 + y_0 + 1 \neq 0$, concluding that $a + b + c = 0$.

Consider lines l_1 and l_2 and view them as equations in a and b: $x_0 a + y_0 b = -c$, $a + x_0 b = -c y_0$. If $x_0 + y_0 + 1 = 0$, then the determinant of the coefficient matrix is $x_0^2 + x_0 + 1$, which is never zero for any real x_0. Solve for a and b in terms of c, we will see that $a = b = c$ and all three lines are the same, contradicting one intersection point.

Now suppose $a + b + c = 0$. Considering the augmented matrix

$$\begin{pmatrix} a & b & -c \\ b & c & -a \\ c & a & -b \end{pmatrix}.$$

By row operations (adding first two rows to the last row), we see the system has a unique solution that gives the intersection point.

Hints and Answers for Chapter 3

3.1 (a) False. Take $A = \left(\begin{smallmatrix} 0 & 1 \\ 0 & 0 \end{smallmatrix}\right)$.

(b) True. If $Ax = \lambda x$, then $0 = A^2x = A(Ax) = A(\lambda x) = \lambda(Ax) = \lambda^2 x$. Since $x \neq 0$, $\lambda = 0$.

(c) False. Take $A = \operatorname{diag}(B, B, B)$, where $B = \left(\begin{smallmatrix} 0 & 1 \\ 0 & 0 \end{smallmatrix}\right)$.

(d) False. Take $A = \operatorname{diag}(0, 1)$.

(e) True.

(f) False.

(g) True.

(h) False.

(i) False in general if $m \neq n$. True if $m = n$.

(j) False.

(k) True.

(l) False.

3.2 It is easy to see that if $AB = BA$ then equality holds. Suppose $(A+B)^2 = A^2+2AB+B^2$. Since $(A+B)^2 = A^2+AB+BA+B^2$ for all square matrices A and B of the same size, we have $AB + BA = 2AB$. It follows that $AB = BA$.

3.3 If $AB = A - B$, one may check that $(A + I)(I - B) = I$. So $I - B$ is the inverse of $A + I$. Thus $(A + I)(I - B) = (I - B)(A + I)$. This implies $AB = BA$. If $AB = A + B$, one can show $(A - I)(B - I) = I$.

3.4 $a = 0$, $b = -2$.

3.5 Scalar matrices kI. Consider $P^{-1}AP = A$ or $PA = AP$ for all P.

3.6 (a) B, C, D. (b) B, C, D over $\mathbb{C}$. C, D over $\mathbb{R}$. (c) D.

3.7 B, C, E, G.

3.8 $a = 0$, b, c are arbitrary.

3.9 Use same elementary row and column operations on A to get B, C.

When $BC = CB$, by computation, $a^2 + b^2 + c^2 - ab - bc - ca = 0$, and by multiplying both sides by $a + b + c$, we have $a^3 + b^3 + c^3 - 3abc = 0$. It is easy to compute that $|\lambda I - A| = \lambda^3 - (a + b + c)\lambda^2$.

3.10 AE_{ij} is the $n \times n$ matrix whose j-th column is the i-th column of A, and 0 elsewhere. $E_{ij}A$ is the $n \times n$ matrix whose i-th row is the j-th row of A, and 0 elsewhere. $E_{ij}AE_{st}$ is the $n \times n$ matrix with the (i,t)-entry a_{js}, and 0 elsewhere.

3.11 $A^2 = -4A$. $A^6 = -2^{10}A$.

3.12 $A = P\,\mathrm{diag}(5,-1)P^{-1}$, where

$$P = \begin{pmatrix} \frac{1}{3} & \frac{1}{3} \\ -\frac{2}{3} & \frac{1}{3} \end{pmatrix}. \quad \text{So} \quad A^{100} = P\begin{pmatrix} 5^{100} & 0 \\ 0 & 1 \end{pmatrix}P^{-1}.$$

3.13 Notice that

$$\begin{pmatrix} 2 & 1 \\ 2 & 3 \end{pmatrix} = \begin{pmatrix} -1 & 1 \\ 1 & 2 \end{pmatrix}\begin{pmatrix} 1 & 0 \\ 0 & 4 \end{pmatrix}\begin{pmatrix} -1 & 1 \\ 1 & 2 \end{pmatrix}^{-1}.$$

It follows that

$$\begin{pmatrix} 2 & 1 \\ 2 & 3 \end{pmatrix}^k = \frac{1}{3}\begin{pmatrix} 2+2^{2k} & 2^{2k}-1 \\ 2^{2k+1}-2 & 2^{2k+1}+1 \end{pmatrix}.$$

$$\begin{pmatrix} \lambda & 1 \\ 0 & \lambda \end{pmatrix}^k = \begin{pmatrix} \lambda^k & k\lambda^{k-1} \\ 0 & \lambda^k \end{pmatrix}.$$

If $k = 2$, then

$$\begin{pmatrix} 0 & 1 & 0 \\ 0 & 0 & 1 \\ 0 & 0 & 0 \end{pmatrix}^k = \begin{pmatrix} 0 & 0 & 1 \\ 0 & 0 & 0 \\ 0 & 0 & 0 \end{pmatrix},$$

and 0 otherwise.

$$\begin{pmatrix} 0 & 1 & 0 \\ 0 & 0 & 1 \\ 1 & 0 & 0 \end{pmatrix}^k = \begin{pmatrix} 0 & 1 & 0 \\ 0 & 0 & 1 \\ 1 & 0 & 0 \end{pmatrix}, \quad \text{when } k = 3m+1,$$

$$\begin{pmatrix} 0 & 1 & 0 \\ 0 & 0 & 1 \\ 1 & 0 & 0 \end{pmatrix}^k = \begin{pmatrix} 0 & 0 & 1 \\ 1 & 0 & 0 \\ 0 & 1 & 0 \end{pmatrix}, \quad \text{when } k = 3m+2,$$

and I_3 otherwise.

3.14 It is easy to see that

$$A^k = \begin{pmatrix} 1 & k \\ 0 & 1 \end{pmatrix} \quad \text{and} \quad PA^kP^{-1} = A,$$

where $P = \left(\begin{smallmatrix} 1 & 1 \\ 0 & k \end{smallmatrix}\right)$. In fact, a more general result can be obtained that if A is an $n \times n$ matrix with all eigenvalues equal to 1, then A^k is similar to A. To see this, it suffices to show the case in which A itself is a Jordan block. Suppose that the Jordan blocks of A^k are $J_1, J_2, \ldots, J_s$, $s \geq 2$, and $P^{-1}A^kP = \operatorname{diag}(J_1, J_2, \ldots, J_s)$. It is easy to see that $r(I - A^k) = n - 1$. However,

$$r(I - P^{-1}A^kP) \leq n - 2,$$

a contradiction. Thus $s = 1$ and A^k is similar to A.

3.15 $A^n = 3^{n-1} \begin{pmatrix} 1 & \frac{1}{2} & \frac{1}{3} \\ 2 & 1 & \frac{2}{3} \\ 3 & \frac{3}{2} & 1 \end{pmatrix}$.

3.16 (a) Use induction on n. If $n = 1$, there is nothing to show. Suppose it is true for $(n-1) \times (n-1)$ matrices. Let λ_1 be an eigenvalue of A and $Au_1 = \lambda_1 u_1$, where u_1 is a nonzero unit vector. Choose $u_2, \ldots, u_n$ such that $U_1 = (u_1, u_2, \ldots, u_n)$ is a unitary matrix. Then $U_1^* A U_1 = \left(\begin{smallmatrix} \lambda_1 & \alpha \\ 0 & A_1 \end{smallmatrix}\right)$, where A_1 is an $(n-1) \times (n-1)$ matrix. The conclusion follows from the induction on A_1.

(b) If $A = P^{-1}BP$, then $f(A) = f(P^{-1}BP) = P^{-1}f(B)P$.

(c) Let U^*AU be as in (a). Then λ^k is an eigenvalue of

$$(U^*AU)^k = U^*A^kU,$$

so λ^k is an eigenvalue of A^k. Similarly, $f(U^*AU) = U^*f(A)U$ and $f(\lambda)$ is an eigenvalue of $f(A)$.

(d) Let $P = \operatorname{diag}(p_1, p_2, \ldots, p_n)$ and $Q = \operatorname{diag}(q_1, q_2, \ldots, q_n)$. Then $AP = QA$ implies $a_{ij}p_j = a_{ij}q_i$. Thus $(p_j - q_i)a_{ij} = 0$, which means either $a_{ij} = 0$ or $p_j = q_i$. Thus $Af(P) = f(Q)A$.

3.17 Consider the n-square matrix A as a linear transformation on $\mathbb{F}$. If A has n linearly independent eigenvectors, say, $u_1, u_2, \ldots, u_n$, corresponding to eigenvalues $\lambda_1, \lambda_2, \ldots, \lambda_n$, not necessarily different. Then these eigenvectors form a basis for $\mathbb{F}^n$ and $(Au_1, Au_2, \ldots, Au_n) = (\lambda_1 u_1, \lambda_2 u_2, \ldots, \lambda_n u_n)$; that is, $AP = \operatorname{diag}(\lambda_1, \lambda_2, \ldots, \lambda_n)P$, where

$P = (u_1, u_2, \dots, u_n)$. So $P^{-1}AP = \operatorname{diag}(\lambda_1, \lambda_2, \dots, \lambda_n)$. Conversely, if A is diagonalizable, i.e., $P^{-1}AP = \operatorname{diag}(\lambda_1, \lambda_2, \dots, \lambda_n)$. Then $AP = \operatorname{diag}(\lambda_1, \lambda_2, \dots, \lambda_n)P$. It follows that the columns of P are the eigenvectors of A. The eigenvalues do not have to be the same.

3.18 There are finite number of λ such that $|\lambda I + A| = 0$. Thus there exist a $\delta > 0$ such that $|\lambda I + A| = 0$ has no solution in $(0, \delta)$.

3.19 By direct computations.

3.20 Let the eigenvalues of A be $\lambda_1, \dots, \lambda_n$. Then the eigenvalues of $p_B(A)$ are $p_B(\lambda_1), \dots, p_B(\lambda_n)$. Thus $p_B(A)$ is invertible if and only if $p_B(\lambda_i) \neq 0$, $i = 1, 2, \dots, n$, i.e., A and B have no common eigenvalues.

3.21 (a) A is singular because $\begin{pmatrix} I & 0 \\ u & 1 \end{pmatrix} A = \begin{pmatrix} B & -Bv \\ 0 & 0 \end{pmatrix}$ has a zero row.

(b) It suffices to show that A has two linearly independent eigenvectors for eigenvalue 0. This is seen by verifying that

$$A \begin{pmatrix} v \\ 1 \end{pmatrix} = 0 \quad \text{and} \quad A \begin{pmatrix} x \\ 0 \end{pmatrix} = 0,$$

where x is a nonzero solution to $Bx = 0$.

(c) Take $P = \begin{pmatrix} I & 0 \\ u & 1 \end{pmatrix}$. Then $PAP^{-1} = \begin{pmatrix} B(I+vu) & -Bv \\ 0 & 0 \end{pmatrix}$ and

$$|\lambda I - A| = \lambda |\lambda I - B(I + vu)|.$$

If λ^2 divides $|\lambda I - A|$, then B or $I + vu$ is singular. Note that $|I + vu| = 1 + uv$. Thus λ^2 divides $|\lambda I - A|$ if and only if B is singular or $uv = -1$.

3.22 Let $f(\lambda) = |A + \lambda(iB)|$. Since $f(1) \neq 0$, $f(\lambda)$ is not identical to zero and $f(\lambda) = 0$ has finite zeros. Let $\beta = -ti$, $t \in \mathbb{R}$, be such a pure imaginary number that $f(\beta) \neq 0$. Then $A + tB$ is nonsingular.

3.23 $|\lambda I - M| = |\lambda^2 I - BA|$ is an even function in λ.

3.24 (a) False.

(b) True.

(c) True.

(d) False.

(e) False.

(f) False.

(g) True.

(h) True.

(i) False.

(j) False.

(k) True.

(l) False.

(m) False.

(n) True.

(o) False.

(p) False. True when λ_i's are distinct.

(q) True.

(r) False.

(s) True.

(t) False.

3.25 (a) Let $u_1, u_2, \ldots, u_n$ be the eigenvectors of A belonging to the eigenvalues $\lambda_1, \lambda_2, \ldots, \lambda_n$, respectively, $\lambda_i \neq \lambda_j$ if $i \neq j$. We first show by induction that $u_1, u_2, \ldots, u_n$ are linearly independent. Let

$$a_1u_1 + a_2u_2 + \cdots + a_nu_n = 0$$

and apply A to the above equation to get

$$a_1\lambda_1u_1 + a_2\lambda_2u_2 + \cdots + a_n\lambda_nu_n = 0.$$

However,

$$a_1\lambda_nu_1 + a_2\lambda_nu_2 + \cdots + a_n\lambda_nu_n = 0.$$

Subtracting,

$$a_1(\lambda_1 - \lambda_n)u_1 + \cdots + a_{n-1}(\lambda_{n-1} - \lambda_n)u_{n-1} = 0.$$

By induction, $u_1, u_2, \ldots, u_{n-1}$ are linearly independent and

$$a_1 = a_2 = \cdots = a_{n-1} = 0$$

since $\lambda_i \neq \lambda_j$ for $i \neq j$, consequently, $a_n = 0$.
Now set $P = (u_1, u_2, \ldots, u_n)$. Then P is an invertible and

$$AP = P\operatorname{diag}(\lambda_1, \lambda_2, \ldots, \lambda_n).$$

It follows that $P^{-1}AP$ is a diagonal matrix.

(b) It is sufficient to show that if A commutes with a diagonal matrix whose diagonal entries are distinct, then A must be diagonal. This can be done by a direct computation.

3.26 $AB = BA^{-1}$ implies $AB^k = B^k A$ for any positive even k. [Note: $AB^k = B^k A^{-1}$ if k is odd.] In particular, $AB^2 = B^2 A$. Since the eigenvalues of A are distinct, B^2 is diagonalizable.

3.27 Let λ_i's be the eigenvalues of A. Then

$$0 = \operatorname{tr}(A^2) - 2\operatorname{tr}(A^3) + \operatorname{tr}(A^4) = \sum_{i=1}^{n} \lambda_i^2 (1-\lambda_i)^2.$$

It follows that $\lambda_i = 0$ or 1, $i = 1, \dots, n$. Since

$$\operatorname{tr}(A^2) = \sum_{i=1}^{n} \lambda_i^2 = c,$$

we see that c is an integer, and c of the λ_i's equal 1, others 0.

If $A^m = A^{m+1}$ for some m, then $A^m = A^k$ for all $k \geq m$. It follows that the eigenvalues of A are all 0's and 1's.

3.28 It suffices to show that all the eigenvalues of A equal zero.

Let $\lambda_1, \lambda_2, \dots, \lambda_n$ be the eigenvalues of A. Then

$$\operatorname{tr} A^k = 0, \quad k = 1, 2, \dots, n,$$

is equivalent to

$$\lambda_1^k + \lambda_2^k + \cdots + \lambda_n^k = 0, \quad k = 1, 2, \dots, n.$$

If all the λ_i's are the same, they must be zero. Otherwise, suppose that $\lambda_{i_1}, \dots, \lambda_{i_m}$ are the distinct nonzero eigenvalues of A. The above equations can be written as

$$l_1 \lambda_{i_1}^k + l_2 \lambda_{i_2}^k + \cdots + l_m \lambda_{i_m}^k = 0, \quad k = 1, 2, \dots, n.$$

Consider the linear equation system

$$\lambda_{i_1}^k x_1 + \lambda_{i_2}^k x_2 + \cdots + \lambda_{i_m}^k x_m = 0, \quad k = 1, 2, \dots, m.$$

An application of the Vandermonde determinant yields that the equation system has only the trivial solution 0. Thus all the eigenvalues of A are zero.

3.29 In the expansion of $(A+B)^k$, there are four kinds of terms: A^k, B^k, $B^m A^{k-m}$, and other terms each has a factor $AB = 0$. Note that

$$\operatorname{tr}(B^m A^{k-m}) = \operatorname{tr}[(B^{m-1}A^{k-m-1})(AB)] = 0.$$

3.30 By induction on k.

3.31 $-1, -1, 5$; $u_1 = (-1,1,0)^t$, $u_2 = (-1,0,1)^t$, $u_3 = (1,1,1)^t$. $P = (u_1, u_2, u_3)$.

3.32 3 is a repeated eigenvalue and its eigenspace has dimension 1. Thus A does not have three linearly independent eigenvalues.

3.33 The eigenvalues of A are 1, 1, -1. To have three linearly independent eigenvectors, the rank of $I - A$ must be 1, which implies $x + y = 0$.

3.34 $a = b = 0$. $T = \begin{pmatrix} \frac{1}{\sqrt{2}} & 0 & \frac{1}{\sqrt{2}} \\ 0 & 1 & 0 \\ -\frac{1}{\sqrt{2}} & 0 & \frac{1}{\sqrt{2}} \end{pmatrix}$.

3.35 Let $Ax = \lambda x$, where $x = (x_1, x_2, \ldots, x_n)^t \neq 0$. Then for each i, $\sum_{i\neq j} a_{ij}x_j = (\lambda - a_{ii})x_i$. Let $|x_{kk}| = \max\{|x_1|, |x_2|, \ldots, |x_n|\} > 0$. Then $(\lambda - a_{kk})x_k = \sum_{j\neq k} a_{kj}x_j$. It follows that

$$|\lambda - a_{kk}| \leq \sum_{j\neq k} |a_{kj}(x_j/x_{kk})| \leq \sum_{j\neq k} |a_{kj}|.$$

3.36 Let U be a unitary matrix such that $T = U^*AU$ is an upper-triangular matrix. Then consider the trace of A^*A.

3.37 (a) $\operatorname{tr} A = \sum_{k=1}^n x_k + i\sum_{k=1}^n y_k$ is real. So $\sum_{k=1}^n y_k = 0$.

(b) Compute

$$\operatorname{tr} A^2 = \sum_{k=1}^n \lambda_k^2 = \sum_{k=1}^n x_k^2 - \sum_{k=1}^n y_k^2 + 2i\sum_{k=1}^n x_k y_k.$$

Since A is real, $\operatorname{tr} A^2$ is real, so $\sum_{k=1}^n x_k y_k = 0$.

(c) See (b).

3.38 If $u_1 + u_2$ were an eigenvector of A, let $A(u_1+u_2) = \mu(u_1+u_2)$. However, $A(u_1+u_2) = \lambda_1 u_1 + \lambda u_2$. Subtracting these equations, we have $0 = (\mu - \lambda_1)u_1 + (\mu - \lambda_2)u_2$. This says that u_1 and u_2 are linearly dependent. This is impossible.

3.39 From $A(u_1, u_2, u_3) = (u_1, 2u_2, 3u_3)$, we have $A = (u_1, 2u_2, 3u_3)P^{-1}$, where $P = (u_1, u_2, u_3)$. Computing the inverse and multiplying gives

$$A = \begin{pmatrix} \frac{7}{3} & 0 & -\frac{2}{3} \\ 0 & \frac{5}{3} & -\frac{2}{3} \\ -\frac{2}{3} & -\frac{2}{3} & 2 \end{pmatrix}.$$

3.40 Consider the matrix A over $\mathbb{C}$. It has two distinct eigenvalues $\pm i$, as does the matrix $\left(\begin{smallmatrix} 0 & -1 \\ 1 & 0 \end{smallmatrix}\right)$. So they are similar over $\mathbb{C}$. Since they are both real matrices, they must be similar over the real too.

3.41 If $c = 0$ and $a \neq d$, then $x_0 = -b/(a-d)$. If $c = 0$ and $a = d$, then A is a scalar matrix aI_2. If $c \neq 0$, then

$$x_0 = [\, (a-d) \pm \sqrt{(a-d)^2 - 4bc}\,]/(2c).$$

3.42 (a) $A^{-1} = \left(\begin{smallmatrix} d & -b \\ -c & a \end{smallmatrix}\right)$.

(b) If $c \neq 0$, then

$$\begin{pmatrix} a & b \\ c & d \end{pmatrix} = \begin{pmatrix} 1 & \frac{a-1}{c} \\ 0 & 1 \end{pmatrix} \begin{pmatrix} 1 & 0 \\ c & 1 \end{pmatrix} \begin{pmatrix} 1 & \frac{d-1}{c} \\ 0 & 1 \end{pmatrix}.$$

If $c = 0$, then $a \neq 0$. Consider

$$\begin{pmatrix} 1 & 0 \\ 1 & 1 \end{pmatrix} \begin{pmatrix} a & b \\ 0 & d \end{pmatrix}.$$

(c) Let λ_1 and λ_2 be the eigenvalues of A. Then

$$\lambda_1\lambda_2 = |A| = 1 \quad \text{and} \quad \lambda_2 = \lambda_1^{-1}.$$

Besides, $\lambda_1 \neq \lambda_2$, since otherwise $|\lambda_1| = |\lambda_2| = 1$, and

$$2 \geq |\lambda_1 + \lambda_2| = |a + d| > 2.$$

Thus A is similar to the diagonal matrix

$$\begin{pmatrix} \lambda_1 & 0 \\ 0 & \lambda_1^{-1} \end{pmatrix}, \quad \lambda_1 \neq 0, \pm 1.$$

(d) If $|a + d| < 2$, then $|\lambda_1 + \lambda_2| = |\lambda_1 + \lambda_1^{-1}| < 2$, so λ_1 is neither real nor purely imaginary.

(e) If $|a+d| = 2$, the possible real eigenvalues of A are 1 and -1. The possible real matrices that A is similar to are

$$I,\ -I,\ T\begin{pmatrix} 1 & 1 \\ 0 & 1 \end{pmatrix}T^{-1},\ T\begin{pmatrix} -1 & 1 \\ 0 & -1 \end{pmatrix}T^{-1},$$

where T is a 2×2 invertible real matrix.

(f) If $|a+d| \neq 2$, then A has two distinct eigenvalues λ_1, λ_2. Thus A is similar to $\operatorname{diag}(\lambda_1, \lambda_2)$. It is easy to check that the matrix

$$\begin{pmatrix} \frac{\lambda_1+\lambda_2}{2} & \frac{\lambda_1-\lambda_2}{2} \\ \frac{\lambda_1-\lambda_2}{2} & \frac{\lambda_1+\lambda_2}{2} \end{pmatrix}$$

has the eigenvalues λ_1, λ_2; therefore, it is also similar to $\operatorname{diag}(\lambda_1, \lambda_2)$.

(g) No. Take $A = \left(\begin{smallmatrix} 1 & 1 \\ 0 & 1 \end{smallmatrix}\right)$.

3.43 The eigenvalues of A and B are 1 and 0. Thus both A and B are diagonalizable and they are similar. A direct computation shows that they are not unitarily similar.

3.44 Let $P^{-1}AP = B$. Then $AP = PB$. Write

$$P = T_1 + iT_2.$$

Then $AT_1 = T_1B$ and $AT_2 = T_2B$. Set

$$T = T_1 + tT_2.$$

Then T is real and invertible for some $t > 0$, and $AT = TB$. Hence A and B are similar over $\mathbb{R}$.

If A and B are two matrices with rational entries and $S^{-1}AS = B$ for some complex matrix S, then $AS = SB$. Consider the matrix equation $AX = XB$. It has either a nonzero solution in $\mathbb{Q}$ or no nonzero solution in $\mathbb{C}$ (as a field is closed under $+, -, \times, \div$).
$M = \left(\begin{smallmatrix} 5 & 1 \\ 3 & 1 \end{smallmatrix}\right)$.

3.45 The eigenvalues of A are $1, 2, 4$, and corresponding eigenvectors are $(1, -1, 1)$, $(1, 0, -1)$, and $(1, 2, 1)$, respectively. They are orthogonal.

3.46 $$A = \begin{pmatrix} \frac{1}{\sqrt{2}} & 0 & \frac{1}{\sqrt{2}} \\ 0 & 1 & 0 \\ \frac{1}{\sqrt{2}} & 0 & -\frac{1}{\sqrt{2}} \end{pmatrix}\begin{pmatrix} \sqrt{2} & 0 \\ 0 & 1 \\ 0 & 0 \end{pmatrix}\begin{pmatrix} 1 & 0 \\ 0 & 1 \end{pmatrix}.$$

3.47 By direct computation: $AT = TA$.

3.48 It may be assumed that A is a Jordan block. Then

$$SA = A^t S \quad \text{or} \quad SAS^{-1} = A^t,$$

where S is the matrix with $(i, n-i+1)$-entry 1, and 0 elsewhere, $i = 1, 2, \ldots, n$. It can also be proved by observing that $\lambda I - A$ and $\lambda I - A^t$ have the same minors. A^* is not similar to A in general because they may have different eigenvalues. A is never similar to $A + I$ because the eigenvalues of $A + I$ are those of A's plus 1.

3.49 If $r(A) < n - 1$, then $\operatorname{adj}(A) = 0$. If $r(A) = n - 1$, then the rank of $\operatorname{adj}(A)$ is 1, the only possible nonzero eigenvalue is $\operatorname{tr}(\operatorname{adj}(A)) = A_{11} + A_{22} + \cdots + A_{nn}$, where A_{ii} is the minor of a_{ii}, $i = 1, 2, \ldots, n$.

3.50 If $A\bar{x} = \sqrt{\lambda}x$, then $\bar{A}x = \sqrt{\lambda}\bar{x}$; So $(\bar{A}A)\bar{x} = \bar{A}(A\bar{x}) = \bar{A}(\sqrt{\lambda}x) = \lambda\bar{x}$; that is, λ is an eigenvalue of $\bar{A}A$, thus an eigenvalue of $A\bar{A}$ since $\lambda \geq 0$. Conversely, assume $A\bar{A}x = \lambda\bar{x}$ with $x \neq 0$. If $\lambda = 0$, let $y = A\bar{x}$. Then $A\bar{y} = A\bar{A}x = \lambda\bar{x} = 0$, as desired. Let $\lambda \neq 0$. If $A\bar{x} = -\sqrt{\lambda}x$, take $y = ix$. If $A\bar{x} \neq -\sqrt{\lambda}x$, take $y = A\bar{x} + \sqrt{\lambda}x$.

3.51 A number of different proofs are given below.

(1) Make use of block matrix techniques: Notice that

$$\begin{pmatrix} I_m & -A \\ 0 & \lambda I_n \end{pmatrix} \begin{pmatrix} \lambda I_m & A \\ B & I_n \end{pmatrix} = \begin{pmatrix} \lambda I_m - AB & 0 \\ \lambda B & \lambda I_n \end{pmatrix}$$

and that

$$\begin{pmatrix} I_m & 0 \\ -B & \lambda I_n \end{pmatrix} \begin{pmatrix} \lambda I_m & A \\ B & I_n \end{pmatrix} = \begin{pmatrix} \lambda I_m & A \\ 0 & \lambda I_n - BA \end{pmatrix}.$$

Take determinants to get

$$\lambda^n |\lambda I_m - AB| = \lambda^m |\lambda I_n - BA|.$$

Thus $|\lambda I_m - AB| = 0$ if and only if $|\lambda I_n - BA| = 0$ when $\lambda \neq 0$. It is immediate that AB and BA have the same nonzero eigenvalues, including multiplicities.

(2) Use elementary operations: Consider the matrix

$$\begin{pmatrix} 0 & 0 \\ B & 0 \end{pmatrix}.$$

Adding the second row premultiplied by A to the first row:

$$\begin{pmatrix} AB & 0 \\ B & 0 \end{pmatrix}.$$

Do the similar operation for columns to get

$$\begin{pmatrix} 0 & 0 \\ B & BA \end{pmatrix}.$$

Write in symbols

$$\begin{pmatrix} I_m & A \\ 0 & I_n \end{pmatrix}\begin{pmatrix} 0 & 0 \\ B & 0 \end{pmatrix} = \begin{pmatrix} AB & 0 \\ B & 0 \end{pmatrix}$$

and

$$\begin{pmatrix} 0 & 0 \\ B & 0 \end{pmatrix}\begin{pmatrix} I_m & A \\ 0 & I_n \end{pmatrix} = \begin{pmatrix} 0 & 0 \\ B & BA \end{pmatrix}.$$

It is immediate that

$$\begin{pmatrix} I_m & A \\ 0 & I_n \end{pmatrix}^{-1}\begin{pmatrix} AB & 0 \\ B & 0 \end{pmatrix}\begin{pmatrix} I_m & A \\ 0 & I_n \end{pmatrix} = \begin{pmatrix} 0 & 0 \\ B & BA \end{pmatrix}.$$

It is readily seen that AB and BA have the same nonzero eigenvalues, counting multiplicities.

(3) Use the argument of continuity: Consider the case where $m = n$. If A is nonsingular, then $BA = A^{-1}(AB)A$. Thus AB and BA are similar, and they have the same eigenvalues.
If A is singular, let δ be such a positive number that $\epsilon I + A$ is nonsingular for every ϵ, $0 < \epsilon < \delta$. Then

$$(\epsilon I + A)B \quad \text{and} \quad B(\epsilon I + A)$$

are similar and have the same characteristic polynomials. Thus

$$|\lambda I_n - (\epsilon I_n + A)B| = |\lambda I_n - B(\epsilon I_n + A)|, \quad 0 < \epsilon < \delta.$$

As both sides are continuous functions of ϵ, letting $\epsilon \to 0$ yields

$$|\lambda I_n - AB| = |\lambda I_n - BA|.$$

It follows that AB and BA have the same eigenvalues.

For the case where $m \neq n$, assume $m < n$ and let

$$A_1 = \begin{pmatrix} A \\ 0 \end{pmatrix}, \quad B_1 = (B, 0)$$

be $n \times n$ matrices. Then

$$A_1 B_1 = \begin{pmatrix} AB & 0 \\ 0 & 0 \end{pmatrix} \text{ and } B_1 A_1 = BA.$$

It follows that $A_1 B_1$ and $B_1 A_1$, consequently AB and BA, have the same nonzero eigenvalues with the same multiplicity.

(4) Treat matrices as operators: We need to show if $\lambda I_m - AB$ is singular, then so is $\lambda I_n - BA$, and vice versa. Assume that $\lambda = 1$. If $I_m - AB$ is invertible, let $X = (I_m - AB)^{-1}$. One may verify

$$(I_n - BA)(I_n + BXA) = I_n.$$

Thus $I_n - BA$ is invertible.

This approach gives no information on multiplicity. Note that

$$|I_m + AB| = |I_n + BA| = \begin{vmatrix} I_n & A \\ B & I_m \end{vmatrix}.$$

3.52 Let $a = (a_1, a_2, \ldots, a_n)^t$ and $e = (1, 1, \ldots, 1)^t$. Denote $B = (a, e)$ and $C = B^t$. Then $A = BC$.

$$|\lambda I - A| = |\lambda I - BC| = \lambda^{n-2}|\lambda I - CB| = \lambda^{n-2}\Big(\lambda - \sum_{i=1}^{n} a_i^2\Big)(\lambda - n).$$

The eigenvalues of A are $0, \ldots, 0$, n, $\sum a_i^2$, all nonnegative.

3.53 $A^2 = 0$ and 0 is the only (repeated) eigenvalue of A. Thus A cannot be similar to a diagonal matrix. The eigenvectors corresponding to 0 are the solutions to $v^t x = 0$. The dimension of the space is $n - 1$.

3.54 It is sufficient to notice that

$$\begin{pmatrix} A & B \\ B & A \end{pmatrix} = S^{-1} \begin{pmatrix} A + B & 0 \\ 0 & A - B \end{pmatrix} S$$

and

$$\begin{pmatrix} A & -B \\ B & A \end{pmatrix} = T^{-1} \begin{pmatrix} A + iB & 0 \\ 0 & A - iB \end{pmatrix} T,$$

where

$$S = \frac{1}{\sqrt{2}} \begin{pmatrix} I & I \\ I & -I \end{pmatrix}, \quad T = \frac{1}{\sqrt{2}} \begin{pmatrix} I & iI \\ -iI & -I \end{pmatrix}.$$

3.55 (a) It follows from Problem 3.51.

(b) By (a),

$$\begin{aligned} \operatorname{tr}(AB)^k &= \operatorname{tr}(AB)(AB)\cdots(AB) \\ &= \operatorname{tr} A(BA)\cdots(BA)B \\ &= \operatorname{tr}(BA)(BA)\cdots(BA) \\ &= \operatorname{tr}(BA)^k. \end{aligned}$$

(c) No, in general. Take

$$A = \begin{pmatrix} 1 & 1 \\ 1 & 0 \end{pmatrix}, \quad B = \begin{pmatrix} 1 & 0 \\ 1 & 0 \end{pmatrix}.$$

(d) If A had an inverse, then $AB - BA = A$ would imply

$$ABA^{-1} - B = I.$$

Thus B is similar to $B + I$. This is impossible.

(e) Write $ABC = A(BC)$, then use (a).

(f) No, in general. Take

$$A = \begin{pmatrix} 0 & 1 \\ 0 & 0 \end{pmatrix}, \quad B = \begin{pmatrix} 1 & 0 \\ 0 & 0 \end{pmatrix}, \quad C = \begin{pmatrix} 0 & 0 \\ 1 & 0 \end{pmatrix}.$$

(g) By (a) and (b).

(h) If A or B is nonsingular, say, A, then $AB = A(BA)A^{-1}$.

(i) No.

3.56 J_n has n eigenvalues $0, \ldots, 0$, and n. The eigenvectors are the solutions to the system $x_1 + x_2 + \cdots + x_n = 0$. K has $2n$ eigenvalues; they are $0, \ldots, 0, -n$ and n. For the eigenvalue $\lambda = 0$, the eigenvectors are the solutions to the systems $x_1 + x_2 + \cdots + x_n = 0$ and $x_{n+1} + x_{n+2} + \cdots + x_{2n} = 0$. The following $2n - 2$ vectors form a basis for the solution space:

$$\alpha_i = (1, 0, \ldots, 0, -1, 0, \ldots, 0), \quad i = 1, 2, \ldots, n-1,$$

where -1 is in the $(i+1)$-position, and

$$\alpha_{n+i} = (0, \ldots, 0, 1, 0, \ldots, 0, -1, 0, \ldots, 0), \quad i = 1, 2, \ldots, n-1,$$

where 1 is in the $(n+1)$-position and -1 is in the $(n+i+1)$-position.

For $\lambda = -n$, an eigenvector is $(1, \ldots, 1, -1, \ldots, -1)$.

For $\lambda = n$, an eigenvector is $(1, \ldots, 1)$.

3.57 $A = J - I$, where J is the matrix all of whose entries are equal to 1. $(J-I)(\frac{1}{n-1}J - I) = I$. Thus $A^{-1} = (J-I)^{-1} = \frac{1}{n-1}J - I$.

3.58 (a) Let $\lambda_1, \ldots, \lambda_s$ be the nonzero eigenvalues of A. Then

$$\operatorname{tr} A = \sum_{i=1}^{s} \lambda_i.$$

Let

$$\hat{\lambda} = \frac{1}{s} \operatorname{tr} A, \quad S = \sum_{i=1}^{s} (\lambda_i - \hat{\lambda})^2 \geq 0.$$

By computation,

$$\begin{aligned} S &= \sum_{i=1}^{s} (\lambda_i - \hat{\lambda})^2 \\ &= \sum_{i=1}^{s} \lambda_i^2 - 2\sum_{i=1}^{s} \lambda_i \hat{\lambda} + \sum_{i=1}^{s} \hat{\lambda}^2 \\ &= \sum_{i=1}^{s} \lambda_i^2 - s\hat{\lambda}^2 \\ &= \operatorname{tr} A^2 - \frac{1}{s}(\operatorname{tr} A)^2. \end{aligned}$$

The desired inequality follows. Equality holds if and only if the nonzero eigenvalues are all the same.

(b) Note that when A is Hermitian, the rank of A is equal to the number of nonzero eigenvalues of A.
If $A^2 = cA$ for some c, then $\lambda_i^2 = c\lambda_i$. It is readily seen that the nonzero eigenvalues are all equal to c.

(c) Let $\lambda_1, \lambda_2, \ldots, \lambda_k$ be the nonzero eigenvalues of A. Then λ_1^2, λ_2^2, $\ldots$, λ_k^2 are nonzero eigenvalues of A^2. By the Cauchy-Schwarz inequality,

$$(\operatorname{tr} A)^2 = (\lambda_1 + \lambda_2 + \cdots + \lambda_k)^2 \leq k(\lambda_1^2 + \cdots + \lambda_k^2) = k \operatorname{tr} A^2.$$

If $(\operatorname{tr} A)^2 > (n-1)\operatorname{tr} A^2$, then k must equal n. Thus $|A| \neq 0$.

3.59 Consider the Jordan blocks of A. The possible eigenvalues of A are 0, 1, and -1.

3.60 We call a "product" a "word". We use the fact that $\operatorname{tr}(AB) = \operatorname{tr}(BA)$ for any square matrices A and B of the same size. First view $S_{5,3}$ as a collection of the words of length 5 with 3 Y's and divide (10 of) them into two groups:

$$XY^2XY,\ Y^2XYX,\ YXYXY,\ XYXY^2,\ YXY^2X$$

and

$$X^2Y^3,\ XY^3X,\ Y^3X^2,\ Y^2X^2Y,\ YX^2Y^2.$$

The words in each group all have the same trace. So

$$\frac{1}{5}\operatorname{tr}(S_{5,3}) = \operatorname{tr}(XY^2XY + X^2Y^3) = \operatorname{tr} X(Y^2XY + XY^3),$$

where $Y^2XY,\ XY^3 \in S_{4,3}$. There are two more elements in $S_{4,3}$: YXY^2 and Y^3X, which have the same trace as YXY^2, XY^3, respectively. (In fact, all the 4 words in $S_{4,3}$ have the same trace.) The conclusion follows at once. One may generalize this to the words of length m with j copies Y and $m-j$ copies of X.

3.61 It is easy to see that the rank of AB is 2 and that

$$(AB)^2 = 9(AB).$$

Thus

$$r(BA) \geq r[A(BA)B] = r(AB)^2 = 2$$

and BA is invertible. However,

$$(BA)^3 = B(AB)^2A = B(9AB)A = 9(BA)^2.$$

Since BA is invertible, it follows that $BA = 9I_2$.

3.62 $(1, 0, -1)$ is an eigenvector of the eigenvalue 3.

$$A = \tfrac{1}{6}\begin{pmatrix} 13 & -2 & 5 \\ -2 & 10 & 2 \\ 5 & 2 & 13 \end{pmatrix}.$$

3.63 $A = \begin{pmatrix} 0 & 1 & 0 \\ 1 & 0 & 0 \\ 0 & 0 & 1 \end{pmatrix}.$

3.64 (a) Direct verification by definition.

(b) $[A, B + C] = A(B + C) - (B + C)A = AB + AC - BA$ $-CA = (AB - BA) + (AC - CA) = [A, B] + [A, C]$.

(c) $[A, B]^* = (AB - BA)^* = B^*A^* - A^*B^* = [B^*, A^*]$.

(d) Note that $P^{-1}[PXP^{-1}, Y]P = [X, P^{-1}YP]$.

(e) $\operatorname{tr}(AB - BA) = \operatorname{tr} AB - \operatorname{tr} BA = 0$.

(f) $\operatorname{tr}(I - [A, B]) = n$. If X is nilpotent, then $\operatorname{tr} X = 0$.

(g) $\operatorname{tr}[A, B] = 0 \neq \operatorname{tr} I = n$.

(h) Take $X = \operatorname{diag}(1, 2, \ldots, n)$ and $Y = (y_{ij})$, where

$$y_{ij} = \begin{cases} \frac{1}{i-j} a_{ij} & \text{if } i \neq j. \\ 0 & \text{if } i = j. \end{cases}$$

Then $A = [X, Y]$. Note that X is Hermitian.

(i) $[A, B] = 0 \Rightarrow AB = BA$. So $A^2B = A(AB) = A(BA) = (AB)A = (AB)A = BA^2$. Inductively for any positive integer p, $A^pB = BA^p$. For the same reason, $A^pB^q = B^qA^p$.

(j) If A is nonsingular, then $AB - BA = A$ implies $ABA^{-1} - B = I$. Taking trace gives $0 = n$. Contradiction.

(k) If A and B are Hermitian, then $[A, B]^* = (AB - BA)^* = B^*A^*$ $-A^*B^* = BA - AB = -(AB - BA) = -[A, B]$. So $[A, B]$ is skew-Hermitian. The other case is similarly proved.

(l) Similar to (k).

(m) See (h). If A is skew-Hermitian, X, Y there are Hermitian.

(n) Let $C = AB - BA$. Then $C^* = (AB)^* - (BA)^* = -C$. So C is skew-Hermitian. Thus iC is Hermitian; all eigenvalues of C are pure imaginary.

(o) It is easy to get from $[A, [A, A^*]] = 0$ that

$$A^2A^* + A^*A^2 = 2AA^*A.$$

Multiplying both sides by A^* from left and taking trace,

$$\operatorname{tr}[(A^*)^2A^2] = \operatorname{tr}[(A^*A)^2],$$

which implies the normality of A (see Chapter 4, Problem 4.91).

(p) By a direct verification.

(q) Let $C = [A, B]$. Show that $C^m = 0$ for some positive integer m. For this, prove by induction that $AB^m - B^m A = mB^{m-1}C$. Let $p(\lambda)$ be the characteristic polynomial of B. Then $p(B) = 0$. Using the above fact, show that $p'(B)C = 0$, $p''(B)C^2 = 0, \ldots,$ $p^{(n)}(B)C^n = 0$. Since $p^{(n)}(B) = n!I$, we have $C^n = 0$. Therefore the eigenvalues of C are necessarily zero.

$[A, B] = [B, A]$ if and only if matrices A and B commute.

3.65 Use the Jordan blocks of A. Or prove as follows. For the fixed λ, let V_1 and V_2 be the solution spaces of $(\lambda I - A)x = 0$ and $(\lambda I - A)^2 x = 0$, respectively. We need to show that there exists an invertible matrix P such that $P^{-1}AP$ is diagonal if and only if $V_1 = V_2$ for every $\lambda \in \mathbb{C}$.

Suppose that A is diagonalizable. We show that $V_1 = V_2$. Let $T^{-1}AT = \operatorname{diag}(\lambda_1, \lambda_2, \ldots, \lambda_n)$, where $\lambda_1, \lambda_2, \ldots, \lambda_n$ be the eigenvalues of A. Then $T^{-1}(\lambda I - A)T = \operatorname{diag}(\lambda - \lambda_1, \lambda - \lambda_2, \ldots, \lambda - \lambda_n)$ and $T^{-1}(\lambda I - A)^2 T = \operatorname{diag}((\lambda - \lambda_1)^2, (\lambda - \lambda_2)^2, \ldots, (\lambda - \lambda_n)^2)$. $\lambda - \lambda_i = 0$ if and only if $\lambda - \lambda_i)^2 = 0$. So $r(\lambda I - A) = r(\lambda I - A)^2$. Since $V_1 \subseteq V_2$, we have $V_1 = V_2$.

Now suppose $V_1 = V_2$. If A is not diagonalizable, we will draw a contradiction. Let J be a Jordan block of A corresponding to an eigenvalue λ_0. If the size of J is more than 1, then $r(\lambda_0 I - J) = r(\lambda_0 I - J)^2 + 1$. Using Jordan form of A, we see that $r(\lambda_0 I - A) > r(\lambda_0 I - A)^2$. It follows that $\dim V_1 < \dim V_2$. A contradiction.

3.66 (a)$\Rightarrow$(b): Obvious.

(b)$\Rightarrow$(c): First note that the linear systems $Ax = 0$ and $A^2 x = 0$ have the same solution space when $r(A) = r(A^2)$. Let $x \in \operatorname{Im} A \cap \operatorname{Ker} A$. Then $Ax = 0$, $x = Ay$ for some y, and $0 = Ax = A(Ay) = A^2 y$; therefore, $0 = Ay$ and $x = 0$.

(c)$\Rightarrow$(d): Choose bases for $\operatorname{Im} A$ and $\operatorname{Ker} A$, they form a basis for $\mathbb{C}^n$. Regard A as a linear transformation on $\mathbb{C}^n$, the matrix representation of A on this basis is of the form $\left(\begin{smallmatrix} D & 0 \\ 0 & 0 \end{smallmatrix}\right)$, where D is invertible.

(d)$\Rightarrow$(a): Notice that

$$A^2 = P \begin{pmatrix} D & 0 \\ 0 & D \end{pmatrix} P^{-1} P \begin{pmatrix} D & 0 \\ 0 & 0 \end{pmatrix} P^{-1} = BA,$$

where $B = P \left(\begin{smallmatrix} D & 0 \\ 0 & D \end{smallmatrix}\right) P^{-1}$ is nonsingular.

3.67 Only A^t always has the same eigenvalues as A, while A^t, $\bar{A}$, A^*, and $(A^*A)^{\frac{1}{2}}$ all have the same singular values as A.

3.68 Let $Au = \lambda u$, $u \neq 0$. We may assume that u is a unit vector. Then $u^*Au = \lambda u^*u = \lambda$. So $\rho \leq \omega$. By the Cauchy-Schwarz inequality, one can show that $\omega \leq \sigma$.

3.69 We show that if $A^*AB = A^*AC$ then $AB = AC$. Notice that $A^*A(B - C) = 0$ implies $(B^* - C^*)A^*A(B - C) = 0$. It follows that $[A(B-C)]^*[A(B-C)] = 0$. Thus $A(B-C) = 0$ and $AB = AC$.

3.70 Since $A^2B = A$, $r(A) = r(A^2B) \leq \min\{r(A^2), r(B)\} \leq r(A)$. So $r(A) = r(A^2) = r(B)$. Thus, the null spaces of A, A^2, and B all have the same dimension. If $Bx = 0$, then $Ax = (A^2B)x = 0$. Hence, the null spaces of A^2 and B are subspaces of the null space A, and they all have to be the same. For any $u \in \mathbb{C}^n$, $(A^2B)(Au) = A(Au)$. So $A^2BAu = A^2u$; that is, $A^2(BAu - u) = 0$, or $BAu - u \in \operatorname{Ker} A^2$. Therefore $B(BAu - u) = 0$, i.e., $B^2Au = Bu$ for all u, or $B^2A = B$.

3.71 (a) $n - 1$. (b) $\operatorname{Im} A = \{\, y \in \mathbb{C}^n \mid y^*x = 0 \,\}$. (c) $\operatorname{Ker} A = \operatorname{Span}\{x\}$.

3.72 The dimension of $M_n(\mathbb{Q})$ over $\mathbb{Q}$ is n^2. Thus

$$I,\ A,\ A^2,\ \ldots,\ A^{n^2}$$

are linearly dependent over $\mathbb{Q}$. Let

$$\frac{a_0}{b_0}I + \frac{a_1}{b_1}A + \frac{a_2}{b_2}A^2 + \cdots + \frac{a_{n^2}}{b_{n^2}}A^{n^2} = 0,$$

where a's and b's are integers and b's are different from 0. Take

$$f(x) = b_0b_1\cdots b_{n^2}\left(\frac{a_0}{b_0} + \frac{a_1}{b_1}x + \frac{a_2}{b_2}x^2 + \cdots + \frac{a_{n^2}}{b_{n^2}}x^{n^2}\right).$$

For $A = \operatorname{diag}(\frac{1}{2}, \frac{2}{3}, \frac{3}{4})$,

$$f(x) = 12\left(x - \frac{1}{2}\right)\left(x - \frac{2}{3}\right)\left(x - \frac{3}{4}\right) = (12x - 6(12x - 8)(12x - 9).$$

3.73 $AX = XB \Rightarrow A^2X = A(AX) = A(XB) = (AX)B = XB^2$. In general, $A^kX = X^kB$ for any positive integer k. Let $p(\lambda) = |\lambda I - A|$ be the characteristic polynomial of A. Then $p(A) = 0$. It follows that $Xp(B) = 0$. Write $p(\lambda) = (\lambda - a_1)(\lambda - a_2)\cdots(\lambda - a_n)$, where $a_1, a_2, \ldots, \alpha_n$ are eigenvalues of A. Since A and B have no common eigenvalues, we see that $p(B)$ is invertible. Thus $X = 0$.

3.74 (a) Let $Av = \lambda v$, $v \neq 0$. Multiplying both sides by $\operatorname{adj}(A)$:

$$\operatorname{adj}(A)Av = \lambda \operatorname{adj}(A)v \quad \text{or} \quad |A|v = \lambda \operatorname{adj}(A)v$$

and $\operatorname{adj}(A)v = \frac{1}{\lambda}|A|v$.

(b) Let $Av = \lambda v$, where $v \neq 0$. If $\lambda \neq 0$, then from the solution of (a), v is an eigenvector of $\operatorname{adj}(A)$.

Suppose $\lambda = 0$. If $r(A) \leq n-2$, then $\operatorname{adj}(A) = 0$ and $\operatorname{adj}(A)v = 0$. If $r(A) = n-1$, then the solution space to $Ax = 0$ has dimension 1 and $\{v\}$ is a basis of the solution space. However, $A(\operatorname{adj}(A)v) = 0$; that is, $\operatorname{adj}(A)v$ is a solution to $Ax = 0$. Thus $\operatorname{adj}(A)v = \mu v$ for some μ.

3.75 (a) Since the eigenvalues of A are all distinct, the eigenvectors of A corresponding to the distinct eigenvalues are linearly independent and they form a basis for $\mathbb{C}^n$. Thus A is diagonalizable. Let $T^{-1}AT = \operatorname{diag}(\lambda_1, \lambda_2, \ldots, \lambda_n)$, where $\lambda_1, \lambda_2, \ldots, \lambda_n$ are the eigenvalues of A. Let $C = T^{-1}BT$. Since $AB = BA$, $\operatorname{diag}(\lambda_1, \lambda_2, \ldots, \lambda_n)C = C\operatorname{diag}(\lambda_1, \lambda_2, \ldots, \lambda_n)$. It follows that $\lambda_i c_{ij} = c_{ij}\lambda_j$ for all i, j. Since $\lambda_i \neq \lambda_j$ when $i \neq j$, we have $c_{ij} = 0$ when $i \neq j$. Thus C is diagonal; that is, $T^{-1}BT$ is diagonal. Now $T^{-1}(AB)T = T^{-1}ATT^{-1}BT$ is also diagonal.

(b) Suppose A and B are diagonalizable. Let T be an invertible matrix such that $T^{-1}AT = \operatorname{diag}(\mu_1 I, \mu_2 I, \ldots, \mu_k I)$, where μ_i are distinct eigenvalues of A, $k \leq n$, and I's are identity matrices of appropriate sizes. Since μ's are different, $AB = BA$ implies that $T^{-1}BT = \operatorname{diag}(B_1, B_2, \ldots, B_k)$, where each B_i is a matrix of the same size as $\mu_i I$. Since B is diagonalizable, all B_i are necessarily diagonalizable. Let $R_i^{-1}B_iR_i$ be diagonal. Set $R = \operatorname{diag}(R_1, R_2, \ldots, R_k)$. Then R is invertible and both $R^{-1}T^{-1}ATR$ and $R^{-1}T^{-1}BTR$ are diagonal.

3.76 (b), (d), (g).

3.77 It is sufficient to notice that $x = (\mathcal{I} - \mathcal{A})x + \mathcal{A}x$.

3.78 It is routine to show that $\mathcal{T}(Y + kZ) = \mathcal{T}(Y) + k\mathcal{T}(Z)$; that is, $\mathcal{T}$ is a linear transformation. When $C = D = 0$, $\mathcal{T}(X) = AXB$. If both A and B are invertible, then $\mathcal{T}\mathcal{A} = \mathcal{A}\mathcal{T} = \mathcal{I}$, where $\mathcal{A}$ is defined by $\mathcal{A}(X) = A^{-1}XB^{-1}$, which is also a linear transformation. Now suppose $\mathcal{T}$ is invertible. Let $\mathcal{T}\mathcal{A} = \mathcal{A}\mathcal{T} = \mathcal{I}$. For the identity matrix I, $I = \mathcal{T}\mathcal{A}(I) = A(\mathcal{A}(I))B$. So A and B must be nonsingular.

3.79 (a) By a direct verification.

(b) $\mathcal{T}(A) = 0$.

(c) Compute $\mathcal{T}^2(X)$, $\mathcal{T}^3(X)$, $\mathcal{T}^4(X), \ldots$, it is readily seen that each term of $\mathcal{T}^{2k}(X)$ contains a factor A^m, $m \geq k$. Thus $\mathcal{T}^{2k} = 0$.

(d) By a direct verification.

(e) Let

$$P^{-1}AP = \operatorname{diag}(\lambda_1, \ldots, \lambda_n)$$

and let P_i be the i-th column of P. Then

$$AP_i = \lambda_i P_i, \quad i = 1, 2, \ldots, n.$$

Let B_{ij} be the matrix having P_i as its j-th column and 0 as other columns. Then $\{B_{ij}\}$ form a basis for $M_n(\mathbb{C})$ and $\mathcal{T}$ has the matrix representation on the basis

$$T = \begin{pmatrix} \lambda_1 I - A^t & 0 & \cdots & 0 \\ 0 & \lambda_2 I - A^t & \cdots & 0 \\ \vdots & \vdots & \ddots & \vdots \\ 0 & 0 & \cdots & \lambda_n I - A^t \end{pmatrix}.$$

It is readily seen that if A is diagonalizable, so is $\mathcal{T}$.

(f) If $\mathcal{T}$ and $\mathcal{L}$ commute, then $\mathcal{T}\mathcal{L}(X) = \mathcal{L}\mathcal{T}(X)$ is equivalent to

$$ABX + XBA = BAX + XAB$$

or

$$(AB - BA)X = X(AB - BA).$$

When A and B commute, $AB - BA = 0$.

$\mathcal{T} = 0$ if and only if A is a scalar matrix.

If $\mathcal{T}$ commutes with $\mathcal{L}$, then $AB - BA$ commutes with any matrix in $M_n(\mathbb{C})$. Thus $AB - BA$ is a scalar matrix. For $\operatorname{tr}(AB - BA) = 0$, we have $AB = BA$.

3.80 (a) Let

$$a_{s+1}\mathcal{A}(\alpha_{s+1}) + \cdots + a_n\mathcal{A}(\alpha_n) = 0.$$

Then

$$\mathcal{A}(a_{s+1}\alpha_{s+1} + \cdots + a_n\alpha_n) = 0,$$

or

$$a_{s+1}\alpha_{s+1} + \cdots + a_n\alpha_n \in \operatorname{Ker} \mathcal{A}.$$

Let

$$a_{s+1}\alpha_{s+1} + \cdots + a_n\alpha_n = a_1\alpha_1 + \cdots + a_s\alpha_s.$$

Then $a_1 = \cdots = a_n = 0$ since $\{\alpha_1, \ldots, \alpha_s, \alpha_{s+1}, \ldots, \alpha_n\}$ is a basis. It follows that $A\alpha_{s+1}, \ldots, A\alpha_n$ are linearly independent.

(b) By (a).

(c) $\{\alpha_1, \ldots, \alpha_s, \alpha_{s+1}, \ldots, \alpha_n\}$ is a basis.

The sum is not a direct sum in general. Consider $\mathcal{A}$ on $\mathbb{R}^2$ defined by $\mathcal{A}(x, y)^t = (x - y, x - y)^t$. It's possible that no β_i falls in $\operatorname{Ker} \mathcal{A}$.

3.81 (a) False. It is always true that $\mathcal{A}(V_1 \cap V_2) \subseteq \mathcal{A}(V_1) \cap \mathcal{A}(V_2)$. But equality does not hold in general. Take V_1 to be the line $y = x$, V_2 to be the x-axis, and $\mathcal{A}$ to be the projection onto x-axis.

(b) True.

(c) True. For every $w \in V_1 + V_2$, let $w = v_1 + v_2$. Then $\mathcal{A}(w) = \mathcal{A}(v_1) + \mathcal{A}(v_2) \in \mathcal{A}(V_1) + \mathcal{A}(V_2)$. So $\mathcal{A}(V_1 + V_2) \subseteq \mathcal{A}(V_1) + \mathcal{A}(V_2)$. However, if $z \in \mathcal{A}(V_1) + \mathcal{A}(V_2)$, then $z = \mathcal{A}(z_1) + \mathcal{A}(z_2) = \mathcal{A}(z_1 + z_2) \in \mathcal{A}(V_1 + V_2)$. So equality holds.

(d) False. Take V_1 to be the line $y = x$, V_2 to be the line $y = -x$, and $\mathcal{A}$ be the projection onto the x-axis.

3.82 The proofs for the equivalence of (a)–(f) are routine. The result does not hold in general when V is of infinite dimension or $\mathcal{A}$ is a linear transformation from V to W. For instance, define $\mathcal{B}$ on $\mathbb{P}[x]$ by

$$\mathcal{B}f(x) = xf(x).$$

Then $\operatorname{Ker} \mathcal{B} = \{0\}$, but $\mathcal{B}$ is not invertible.

3.83 (a) Consider A as a linear transformation on $\mathbb{C}^n$. The vectors v, $A(v)$, $A^2(v), \ldots, A^{n-1}(v)$ form a basis for $\mathbb{C}^n$. The matrix presentation of the linear transformation under this basis has a submatrix I_{n-1} on the upper-right corner. Thus for any eigenvalue λ, $r(\lambda I - A) = n - 1$. So $\dim \operatorname{Ker}(\lambda I - A) = 1$, and the eigenvectors belonging to λ are multiple of each other.

(b) Let $u_1, u_2, \ldots, u_n$ be eigenvectors, respectively, corresponding to the distinct eigenvalues $\lambda_1, \lambda_2, \ldots, \lambda_n$ of $\mathcal{A}$. Let $u = u_1 + u_2 + \cdots + u_n$. Then $\mathcal{A}(u) = \lambda_1 u_1 + \lambda_2 u_2 + \cdots + \lambda_n u_n$, $\mathcal{A}^2(u) = \lambda_1^2 u_1 +$

$\lambda_2^2 u_2 + \cdots + \lambda_n^2 u_n, \ldots, \mathcal{A}^{n-1}(u) = \lambda_1^{n-1} u_1 + \lambda_2^{n-1} u_2 + \cdots + \lambda_n^{n-1} u_n$. The coefficient matrix of $u, \mathcal{A}(u), \ldots, \mathcal{A}^{n-1}(u)$ under the basis $u_1, u_2, \ldots, u_n$ is a Vandermonde matrix. This matrix is nonsingular for distinct $\lambda_1, \lambda_2, \ldots, \lambda_n$. So $u, \mathcal{A}(u), \mathcal{A}^2(u), \ldots, A^{n-1}(u)$ are linearly independent.

3.84 Let

$$a_1 x + a_2 \mathcal{A}(x) + \cdots + a_n \mathcal{A}^{n-1}(x) = 0.$$

Applying $\mathcal{A}^k$, $k = n-1, n-2, \ldots, 1$, to both sides of the equation,

$$a_1 = a_2 = \cdots = a_n = 0.$$

The eigenvalues of $\mathcal{A}$ are all zero. The matrix of $\mathcal{A}$ under the basis is the matrix with all $(i, i+1)$-entries 1 and 0 elsewhere.

3.85 Use matrix representations. For matrices A and B, if $AB = I$, then $BA = I$. It is not true for infinite dimensional spaces.

Consider $\mathbb{P}[x]$ with $\mathcal{A}$ and $\mathcal{B}$ defined as

$$\mathcal{A}f(x) = a_1 + a_2 x + \cdots + a_n x^{n-1}$$

and

$$\mathcal{B}f(x) = xf(x),$$

where $f(x) = a_0 + a_1 x + \cdots + a_n x^n$.

3.86 Let $k_1 u_1 + k_2 u_2 + \cdots + k_n u_n = 0$. Applying the linear transformation $\mathcal{A}$ to it yields $k_1 \lambda_1 u_1 + k_2 \lambda_2 u_2 + \cdots + k_n \lambda_n u_n = 0$. Multiplying $k_1 u_1 + k_2 u_2 + \cdots + k_n u_n = 0$ by λ_n, then subtracting, we see $k_1(\lambda_1 - \lambda_n) u_1 + k_2(\lambda_2 - \lambda_n) u_2 + \cdots + k_{n-1}(\lambda_{n-1} - \lambda_n) u_{n-1} = 0$. By induction, u_1, $u_2, \ldots, u_{n-1}$ are linearly independent. So all k_i must be 0, $i = 1, 2, \ldots, n-1$. It follows that k_n has to be 0 too.

3.87 (a) False.

(b) False.

(c) True.

(d) False.

(e) True.

(f) False.

(g) True.

(h) True.

(i) False.

(j) True.

(k) True.

(l) False. One direction is right.

(m) True.

(n) False. One direction is right.

3.88 (a) False. Consider $\mathcal{A} = 0$.

(b) True. If the vectors $\alpha_1, \alpha_2, \ldots, \alpha_n$ are linearly dependent, then there exist $k_1, k_2, \ldots, k_n$, not all zero, such that

$$k_1\alpha_1 + k_2\alpha_2 + \cdots + k_n\alpha_n = 0,$$

which leads to

$$k_1\mathcal{A}\alpha_1 + k_2\mathcal{A}\alpha_2 + \cdots + k_n\mathcal{A}\alpha_n = 0,$$

a contradiction.

3.89 (a) $\mathcal{A}(\alpha_1, \alpha_2, \alpha_3) = (\alpha_1, \alpha_2, \alpha_3)A$, where $A = \begin{pmatrix} 1 & 1 & 1 \\ 0 & 1 & 1 \\ 0 & 0 & 1 \end{pmatrix}$. Since A is invertible, $\mathcal{A}$ is invertible.

(b) $A^{-1} = \begin{pmatrix} 1 & -1 & 0 \\ 0 & 1 & -1 \\ 0 & 0 & 1 \end{pmatrix}$.

So $\mathcal{A}^{-1}(\alpha_1) = \alpha_1$, $\mathcal{A}^{-1}(\alpha_2) = \alpha_2 - \alpha_1$, $\mathcal{A}^{-1}(\alpha_3) = \alpha_3 - \alpha_2$.

(c) The matrix of $2\mathcal{A} - \mathcal{A}^{-1}$ under the basis $\{\alpha_1, \alpha_2, \alpha_3\}$ is

$2A - A^{-1} = \begin{pmatrix} 1 & 3 & 2 \\ 0 & 1 & 3 \\ 0 & 0 & 1 \end{pmatrix}$.

3.90 They are all $p(\lambda) = \lambda^3$.

3.91 For convenience, denote

$$u_1 = \begin{pmatrix} 1 \\ 0 \\ 1 \end{pmatrix}, \ u_2 = \begin{pmatrix} 1 \\ -1 \\ 1 \end{pmatrix}, \ u_3 = \begin{pmatrix} -2 \\ 7 \\ -1 \end{pmatrix}.$$

To find $\operatorname{Im}\mathcal{A}$, apply row operations to $(\mathcal{A}u_1, \mathcal{A}u_2, \mathcal{A}u_3)$:

$$\begin{pmatrix} 2 & 3 & 2 \\ 3 & 0 & 3 \\ -1 & -2 & -1 \end{pmatrix} \longrightarrow \begin{pmatrix} 1 & 0 & 1 \\ 0 & 1 & 0 \\ 0 & 0 & 0 \end{pmatrix}.$$

Thus $\{\mathcal{A}u_1, \mathcal{A}u_2\}$ is a basis for $\operatorname{Im}\mathcal{A}$ and

$$\operatorname{Im}\mathcal{A} = \operatorname{Span}\{\mathcal{A}u_1, \mathcal{A}u_2\}.$$

To find an equation for $\mathcal{A}$, let

$$x = (x_1, x_2, x_3)^t = y_1u_1 + y_2u_2 + y_3u_3.$$

Then

$$\begin{pmatrix} 1 & 1 & -2 \\ 0 & -1 & 7 \\ 1 & 1 & -1 \end{pmatrix} \begin{pmatrix} y_1 \\ y_2 \\ y_3 \end{pmatrix} = \begin{pmatrix} x_1 \\ x_2 \\ x_3 \end{pmatrix}.$$

Denote by B the 3×3 matrix on the left-hand side. Then $By = x$, where $y = (y_1, y_2, y_3)^t$, and $y = B^{-1}x$, where

$$B^{-1} = \begin{pmatrix} 6 & 1 & -5 \\ -7 & -1 & 7 \\ -1 & 0 & 1 \end{pmatrix}.$$

Thus

$$\begin{aligned} \mathcal{A}(x) &= y_1\mathcal{A}u_1 + y_2\mathcal{A}u_2 + y_3\mathcal{A}u_3 \\ &= (\mathcal{A}u_1, \mathcal{A}u_2, \mathcal{A}u_3)y \\ &= (\mathcal{A}u_1, \mathcal{A}u_2, \mathcal{A}u_3)B^{-1}x \\ &= \begin{pmatrix} -11 & -1 & 13 \\ 15 & 3 & -12 \\ 9 & 1 & -10 \end{pmatrix} \begin{pmatrix} x_1 \\ x_2 \\ x_3 \end{pmatrix}. \end{aligned}$$

3.92 (a) Consider $Ax = 0$ to get a basis for $\operatorname{Ker} A$:

$$\alpha_1 = \left(-2, -\frac{3}{2}, 1, 0\right)^t, \quad \alpha_2 = (-1, -2, 0, 1)^t.$$

Let

$$\beta_1 = -2\epsilon_1 - \frac{3}{2}\epsilon_2 + \epsilon_3$$

and

$$\beta_2 = -\epsilon_1 - 2\epsilon_2 + \epsilon_4.$$

Then

$$\operatorname{Ker} \mathcal{A} = \operatorname{Span}\{\beta_1, \beta_2\}.$$

(b) To find $\operatorname{Im} \mathcal{A}$, apply row operations to $(\mathcal{A}\epsilon_1, \mathcal{A}\epsilon_2, \mathcal{A}\epsilon_3, \mathcal{A}\epsilon_4)$ to get

$$\operatorname{Im} \mathcal{A} = \operatorname{Span}\{\mathcal{A}\epsilon_1, \mathcal{A}\epsilon_2\}.$$

(c) $\{\beta_1, \beta_2, \epsilon_1, \epsilon_2\}$ serves as a basis for V. The matrix representation of $\mathcal{A}$ under this basis is

$$\begin{pmatrix} 5 & 2 & 0 & 0 \\ \frac{9}{2} & 1 & 0 & 0 \\ 1 & 2 & 0 & 0 \\ 2 & -2 & 0 & 0 \end{pmatrix}.$$

3.93 (a) The matrix of $\mathcal{A} + \mathcal{B}$ under β_1, β_2 is $\left(\begin{smallmatrix} 8 & 9 \\ \frac{4}{3} & 3 \end{smallmatrix}\right)$.

(b) The matrix of $\mathcal{AB}$ under α_1, α_2 is $\left(\begin{smallmatrix} 7 & 8 \\ 13 & 14 \end{smallmatrix}\right)$.

(c) The coordinate of $\mathcal{A}(u)$ under α_1, α_2 is $(3, 5)$.

(d) The coordinate of $\mathcal{B}(u)$ under β_1, β_2 is $(9, 6)$.

3.94 Take a basis for W, then extend it to a basis for V.

3.95 (a) Apply $\mathcal{A}^{-1}$ to both sides of $\mathcal{A}(W) \subseteq W$ to get

$$W \subseteq \mathcal{A}^{-1}(W).$$

However,

$$\dim \mathcal{A}^{-1}(W) \leq \dim W.$$

Therefore,

$$W = \mathcal{A}^{-1}(W).$$

(b) No, in general.

3.96 With matrix A, we see that $\mathcal{A}(\alpha_1) = 2\alpha_1$ and $\mathcal{A}(\alpha_2) = \alpha_1 + 2\alpha_2$. Let $k\alpha_1 \in W_1$. Then $\mathcal{A}(k\alpha_1) = k\mathcal{A}(\alpha_1) = 2k\alpha_1 \in W_1$. So W_1 is invariant under $\mathcal{A}$. If W_2 is an invariant subspace such that $\mathbb{R}^2 = W_1 \oplus W_2$, then the dimension of W_2 is 1. Let $\alpha_2 = p\alpha_1 + w_2$, where $w_2 \in W_2$, and let $\mathcal{A}(w_2) = qw_2$. From $\mathcal{A}(\alpha_2) = \alpha_1 + 2\alpha_2$, we have $\alpha_1 + 2\alpha_2 = 2p\alpha_1 + qw_2$. Subtracting $2\alpha_2 = 2p\alpha_1 + 2w_2$, we have $\alpha_1 = (q-2)w_2$, which is in both W_1 and W_2. But $W_1 \cap W_2 = \{0\}$. A contradiction.

3.97 (a) It is routine to show that $\mathcal{A}$ is a linear transformation on $M_2(\mathbb{R})$.

(b) The matrix of $\mathcal{A}$ under the basis is $\begin{pmatrix} 1 & 0 & -1 & 0 \\ 0 & 1 & 0 & -1 \\ -1 & 0 & 1 & 0 \\ 0 & -1 & 0 & 1 \end{pmatrix}$.

(c) $\dim(\operatorname{Im}\mathcal{A}) = 2$. $\left(\begin{smallmatrix} -1 & 0 \\ 1 & 0 \end{smallmatrix}\right)$ and $\left(\begin{smallmatrix} 0 & -1 \\ 0 & 1 \end{smallmatrix}\right)$ form a basis.

(d) $\dim(\operatorname{Ker} A) = 2$. $\left(\begin{smallmatrix} 1 & 0 \\ 1 & 0 \end{smallmatrix}\right)$ and $\left(\begin{smallmatrix} 0 & 1 \\ 0 & 1 \end{smallmatrix}\right)$ form a basis.

3.98 With $\mathcal{A}^2 = \mathcal{A}$ and $\mathcal{B}^2 = \mathcal{B}$, one may show that $\mathcal{A}(\mathcal{A}-\mathcal{B})^2 = \mathcal{A}-\mathcal{A}\mathcal{B}\mathcal{A}$. Similarly, $(\mathcal{A}-\mathcal{B})^2\mathcal{A} = \mathcal{A}-\mathcal{A}\mathcal{B}\mathcal{A}$. So $\mathcal{A}$ commutes with $(\mathcal{A}-\mathcal{B})^2$. For the second part of the problem, it is sufficient to notice that

$$(\mathcal{I}-\mathcal{A}-\mathcal{B})^2 = [\mathcal{I}-(\mathcal{A}+\mathcal{B})]^2 = \mathcal{I}-2\mathcal{A}-2\mathcal{B}+(\mathcal{A}+\mathcal{B})^2.$$

3.99 $\dim \operatorname{Im}\mathcal{A} = 2$ and ϵ_1, ϵ_2 for a basis. $\dim \operatorname{Ker}\mathcal{A} = 2$ and $\zeta_1 = \epsilon_1 - \epsilon_2$ and $\zeta_2 = \epsilon_1 - \epsilon_3$ form a basis. $\dim(\operatorname{Im}\mathcal{A} + \operatorname{Ker}\mathcal{A}) = 3$ and $\epsilon_1, \epsilon_2, \epsilon_3$ form a basis. $\dim(\operatorname{Im}\mathcal{A} \cap \operatorname{Ker}\mathcal{A}) = 1$ and $\zeta_3 = \epsilon_1 - \epsilon_2$ is a basis.

3.100 (a) It is routine.

(b) The line $y = x$.

(c) $\operatorname{Ker}\mathcal{B}$ and $\operatorname{Im}\mathcal{B}$ are both the line $y = x$.

(d) $\operatorname{Ker}\mathcal{B}$ and $\operatorname{Im}\mathcal{B}$ have nonzero elements in common.

3.101 (a) By definition.

(b) $\mathcal{A}\mathcal{B}(x_1, x_2, \ldots, x_n) = (0, x_n, x_1, x_2, \ldots, x_{n-2})$.
$\mathcal{B}\mathcal{A}(x_1, x_2, \ldots, x_n) = (x_{n-1}, 0, x_1, x_2, \ldots, x_{n-1})$.
$\mathcal{A}^n = 0$ and $\mathcal{B}^n = \mathcal{I}$.

(c) Under the standard basis (column vectors) $e_1, e_2, \ldots, e_n$, $A = (e_2, e_3, \ldots, e_{n-1}, 0)$ and $B = (e_2, e_3, \ldots, e_{n-1}, e_1)$.

(d) The dimensions of $\operatorname{Ker}\mathcal{A}$ and $\operatorname{Ker}\mathcal{B}$ are 1 and 0, respectively.

3.102 Let $\{\gamma_1, \ldots, \gamma_r\}$ be a basis for $\operatorname{Ker}\mathcal{A}$. To show that V is the direct sum of the subspace spanned by $\beta_1, \ldots, \beta_m$ and $\operatorname{Ker}\mathcal{A}$, we show that $\{\beta_1, \ldots, \beta_m, \gamma_1, \ldots, \gamma_r\}$ is a basis for V.

Let $v \in V$. Then $\mathcal{A}(v) \in \operatorname{Im}\mathcal{A}$. Writing

$$\mathcal{A}(v) = a_1\alpha_1 + \cdots + a_m\alpha_m$$

and replacing α_i by $\mathcal{A}(\beta_i)$, we have

$$\mathcal{A}(v) = a_1\mathcal{A}(\beta_1) + \cdots + a_m\mathcal{A}(\beta_m)$$

and

$$\mathcal{A}(v - a_1\beta_1 - \cdots - a_m\beta_m) = 0.$$

Thus

$$v - a_1\beta_1 - \cdots - a_m\beta_m \in \operatorname{Ker}\mathcal{A}.$$

Let

$$v - a_1\beta_1 - \cdots - a_m\beta_m = b_1\gamma_1 + \cdots + b_r\gamma_r.$$

Then

$$v = a_1\beta_1 + \cdots + a_m\beta_m + b_1\gamma_1 + \cdots + b_r\gamma_r.$$

Therefore

$$V = \operatorname{Span}\{\beta_1, \ldots, \beta_m\} + \operatorname{Ker}\mathcal{A}.$$

Now show that $\beta_1, \ldots, \beta_m, \gamma_1, \ldots, \gamma_r$ are linearly independent. Let

$$c_1\beta_1 + \cdots + c_m\beta_m + d_1\gamma_1 + \cdots + d_r\gamma_r = 0.$$

Applying $\mathcal{A}$ to both sides of the above identity gives

$$c_1\mathcal{A}(\beta_1) + \cdots + c_m\mathcal{A}(\beta_m) + d_1\mathcal{A}(\gamma_1) + \cdots + d_r\mathcal{A}(\gamma_r) = 0,$$

that is,

$$c_1\alpha_1 + \cdots + c_m\alpha_m = 0.$$

Thus $c_1 = \cdots = c_m = 0$ due to the independence of $\alpha_1, \ldots, \alpha_m$. So

$$d_1\gamma_1 + \cdots + d_r\gamma_r = 0,$$

and $d_1 = \cdots = d_r = 0$ for the similar reason. The conclusion follows.

3.103 If $V = \operatorname{Im}\mathcal{A} \oplus \operatorname{Ker}\mathcal{A}$, we show that $\operatorname{Im}\mathcal{A}^2 = \operatorname{Im}\mathcal{A}$. Obviously, $\operatorname{Im}\mathcal{A}^2 \subseteq \operatorname{Im}\mathcal{A}$. Let $u \in \operatorname{Im}\mathcal{A}$. Then $u = \mathcal{A}v$ for some $v \in V$. Write $v = w_1 + w_2$, where $w_1 \in \operatorname{Im}\mathcal{A}$ and $w_2 \in \operatorname{Ker}\mathcal{A}$. Let $w_1 = \mathcal{A}z_1$. Then $u = \mathcal{A}v = \mathcal{A}(w_1) + \mathcal{A}(w_2) = \mathcal{A}(w_1) = \mathcal{A}^2(z_1) \in \operatorname{Im}\mathcal{A}^2$. Therefore, $\operatorname{Im}\mathcal{A}^2 = \operatorname{Im}\mathcal{A}$ and $r(\mathcal{A}^2) = \dim(\operatorname{Im}\mathcal{A}^2) = \dim(\operatorname{Im}\mathcal{A}) = r(\mathcal{A})$.

3.104 (a), (b), (c), (g) are easy to check. (k), (l), (m) follow from (j).

(d) $(\mathcal{A} + \mathcal{I})^{-1} = -\frac{1}{2}\mathcal{A} + \mathcal{I}$.

(e) Note that if $x \in \operatorname{Ker}\mathcal{A}$, then $x = x - \mathcal{A}x$.

(f) First note that $V = \operatorname{Im} \mathcal{A} + \operatorname{Ker} \mathcal{A}$, since

$$v = \mathcal{A}v + (v - \mathcal{A}v).$$

Now let

$$z \in \operatorname{Im} \mathcal{A} \cap \operatorname{Ker} \mathcal{A} \quad \text{and} \quad z = \mathcal{A}y.$$

Set $\mathcal{A}y = x - \mathcal{A}x$ for some $x \in V$ by (e). Then

$$x = \mathcal{A}y + \mathcal{A}x.$$

Applying $\mathcal{A}$ to both sides results in

$$\mathcal{A}x = \mathcal{A}^2 y + \mathcal{A}^2 x = \mathcal{A}y + \mathcal{A}x.$$

Thus $z = \mathcal{A}y = 0$ and $\operatorname{Im} \mathcal{A} \cap \operatorname{Ker} \mathcal{A} = \{0\}$.

(h) Let $\mathcal{B}$ be the linear transformation on V such that

$$\mathcal{B}x = x, \quad x \in M \quad \text{and} \quad \mathcal{B}y = 0, \quad y \in L.$$

Such a $\mathcal{B}$ is uniquely determined by M and L.

(i) If $\mathcal{A}x = \lambda x$, $x \neq 0$, then

$$\mathcal{A}^2 x = \lambda \mathcal{A}x = \lambda^2 x.$$

Since $\mathcal{A}^2 x = \mathcal{A}x$,

$$(\lambda^2 - \lambda)x = 0.$$

Thus $\lambda = 0$ or $\lambda = 1$.

(j) By (f), take a basis for $\operatorname{Im} \mathcal{A}$ and a basis for $\operatorname{Ker} \mathcal{A}$ to form a basis for V. Then the matrix representation of $\mathcal{A}$ under the basis is of the desired form.

3.105 (a) If $\mathcal{A}x = \lambda x$, then $\mathcal{A}(\mathcal{B}x) = \mathcal{B}(\mathcal{A}x) = \lambda(\mathcal{B}x)$, thus $\mathcal{B}x \in V_\lambda$.

(b) If $x \in \operatorname{Ker} \mathcal{A}$, then $\mathcal{A}x = 0$. Note that $\mathcal{A}(\mathcal{B}x) = \mathcal{B}(\mathcal{A}x) = 0$, thus $\mathcal{B}x \in \operatorname{Ker} \mathcal{A}$, and $\operatorname{Ker} \mathcal{A}$ is invariant under $\mathcal{B}$. Similarly, $\operatorname{Im} \mathcal{A}$ is also invariant under $\mathcal{B}$.

(c) Let $\mathcal{B}_\lambda$ be the restriction of $\mathcal{B}$ on V_λ; that is, $\mathcal{B}_\lambda(v) = \mathcal{B}(v)$, $v \in V_\lambda$. $\mathcal{B}_\lambda$ has an eigenvalue in $\mathbb{C}$ and an eigenvector in V_λ.

(d) By induction on dimension. Take v to be a common eigenvector of $\mathcal{A}$ and $\mathcal{B}$. Let W be a subspace such that $V = \operatorname{Span}\{v\} \oplus W$. Let $\mathcal{A}_1$ and $\mathcal{B}_1$ be the restrictions of $\mathcal{A}$ and $\mathcal{B}$ on W, respectively. Then $\mathcal{A}_1$ and $\mathcal{B}_1$ commute. Now apply induction hypothesis.

When $\mathbb{C}$ is replaced by $\mathbb{R}$, (a) and (b) remain true.

3.106 (a) By definition.

(b) 0.

(c) The matrices of $\mathcal{D}$ under the bases are, respectively,

$$D_1 = (0, e_1, 2e_2, \ldots, (n-1)e_{n-1}), \quad D_2 = (0, e_1, e_2, \ldots, e_{n-1}),$$

where the e's are the vectors in the standard basis of $\mathbb{R}^n$.

(d) No, since all eigenvalues are $\mathcal{D}$ are 0.

3.107 (a) Note that

$$(e^{\lambda x})' = \lambda e^{\lambda x} \quad \text{and} \quad (e^{\lambda x})'' = \lambda^2 e^{\lambda x}.$$

(b) $ce^{\lambda x}$ is an eigenvector of $\mathcal{D}_2$ belonging to the eigenvalue λ^2. For any positive number λ, it is easy to see that

$$\mathcal{D}_2(e^{\sqrt{\lambda}x}) = (\sqrt{\lambda}\,)^2(e^{\sqrt{\lambda}x}) = \lambda e^{\sqrt{\lambda}x}.$$

Hence λ is an eigenvalue of $\mathcal{D}_2$.

3.108 (a) Let $p,\ q \in \mathbb{P}_n[x]$. Then

$$\begin{aligned}
\mathcal{A}((p+kq)(x)) &= \mathcal{A}((p(x)+kq(x)) \\
&= x(p(x)+kq(x)' - (p(x)+kq(x)) \\
&= xp'(x) + xkq'(x) - p(x) - kq(x) \\
&= \mathcal{A}(p(x)) + k\mathcal{A}(q(x)).
\end{aligned}$$

So $\mathcal{A}$ is a linear transformation on $\mathbb{P}_n[x]$.

(b) $\operatorname{Ker}\mathcal{A} = \{\, kx \mid k \in \mathbb{R}\,\}$.
$\operatorname{Im}\mathcal{A} = \{\, a_0 + a_2x^2 + \cdots + a_{n-1}x^{n-1} \mid a_0, a_2, \ldots, a_{n-1} \in \mathbb{R}\,\}$.

(c) By (b).

3.109 (a) Let W be an invariant subspace of V under $\mathcal{A}$. Then W is invariant under $(\mathcal{A} - \lambda\mathcal{I})^i$ for $i = 1, 2, \ldots, n$. Observe that

$$\begin{aligned}
u_2 &= (\mathcal{A} - \lambda\mathcal{I})u_1 \\
u_3 &= (\mathcal{A} - \lambda\mathcal{I})u_2 \\
\vdots & \qquad \vdots \\
u_n &= (\mathcal{A} - \lambda\mathcal{I})u_{n-1}
\end{aligned}$$

or

$$u_i = (\mathcal{A} - \lambda\mathcal{I})^{i-1}u_1, \quad i = 1, 2, \ldots, n.$$

Since W is invariant under

$$(\mathcal{A} - \lambda\mathcal{I})^{i-1}, \quad i = 1, 2, \ldots, n,$$

if $u_1 \in W$, then $u_2, \ldots, u_n \in W$ and $W = V$.

(b) Let $x \in W$, $x \neq 0$, and let

$$x = a_i u_i + \cdots + a_n u_n,$$

where $i \geq 1$ and $a_i \neq 0$. If $i = n$, it is trivial. Suppose $i < n$. Since W is invariant under $\mathcal{A}$, consequently under $(\mathcal{A} - \lambda\mathcal{I})^{n-i}$,

$$(\mathcal{A} - \lambda\mathcal{I})^{n-i} x \in W, \quad (\mathcal{A} - \lambda\mathcal{I}) u_n = 0,$$

and

$$\begin{aligned} (\mathcal{A} - \lambda\mathcal{I})^{n-i} x &= (\mathcal{A} - \lambda\mathcal{I})^{n-i}(a_i u_i + \cdots + a_n u_n) \\ &= a_i (\mathcal{A} - \lambda\mathcal{I})^{n-1} u_1 + 0 \\ &= a_i u_n \in W. \end{aligned}$$

It turns out that $u_n \in W$.

(c) Observe that for $k = n - i + 1, \ldots, n - 1$,

$$(\mathcal{A} - \lambda\mathcal{I}) u_k = u_{k+1} \in V_i$$

and

$$\mathcal{A} u_n = \lambda u_n \in V_i.$$

Thus each V_i is invariant under $\mathcal{A} - \lambda\mathcal{I}$. Write

$$\mathcal{A} = (\mathcal{A} - \lambda\mathcal{I}) + \lambda\mathcal{I}.$$

It follows that V_i is invariant under $\mathcal{A}$.

To show that

$$(\mathcal{A} - \lambda\mathcal{I})^i x = 0 \quad \Leftrightarrow \quad x \in V_i,$$

let x be a linear combination of $u_1, u_2, \ldots, u_n$

$$x = x_1 u_1 + x_2 u_2 + \cdots + x_n u_n.$$

If $(\mathcal{A} - \lambda\mathcal{I})^i x = 0$, applying $(\mathcal{A} - \lambda\mathcal{I})^i$ to both sides results in

$$0 = x_1 u_{i+1} + \cdots + x_{n-i} u_n.$$

Thus $x_1 = x_2 = \cdots = x_{n-i} = 0$, and $x \in V_i$.

The other direction is immediate by observing that

$$(\mathcal{A} - \lambda\mathcal{I})^i u_k = 0, \quad k \geq n - i + 1.$$

(d) Let W be an invariant subspace of $\mathcal{A}$ with dimension m. Then there must exist an $x \in W$ such that

$$x = a_i u_i + \cdots + a_n u_n, \quad a_i \neq 0, \ i \leq n - m + 1.$$

Applying

$$(\mathcal{A} - \lambda \mathcal{I})^{n-k}, \quad k = i, i+1, \ldots, i+m-1,$$

to both sides consecutively, we have

$$u_n, u_{n-1}, \ldots, u_{n-m+1} \in V_m.$$

Thus $W = V_m$ since they have the same dimension.

(e) Note that an eigenspace is invariant.

(f) If $V = W \oplus U$ and W and U are nontrivial, then $u_n \in W$ and $u_n \in U$. This is impossible.

For the matrix S such that $SAS^{-1} = A^t$, take S to be the matrix with all $(i, n-i+1)$-entries 1, $i = 1, 2, \ldots, n$, and 0 elsewhere.

3.110 If λ_0 is an eigenvalue of $\mathcal{L}$,

$$\mathcal{L}(X_0) = \lambda_0 X_0, \quad \text{for some} \quad X_0 \neq 0,$$

then

$$AX_0 = \lambda_0 X_0.$$

If x_0 is a nonzero column vector of X_0, then $Ax_0 = \lambda_0 x_0$ and λ_0 is an eigenvalue of A. Conversely, if x_0 is an eigenvector of A belonging to λ_0, let X_0 be the matrix with all column vectors x_0. Then

$$\mathcal{L}(X_0) = AX_0 = \lambda_0 X_0,$$

and λ_0 is an eigenvalue of $\mathcal{L}$.

The characteristic polynomial of $\mathcal{L}$ is the n-th power of the characteristic polynomial of A. To see this, consider the basis of $M_n(\mathbb{C})$

$$\{E_{11}, E_{21}, \ldots, E_{n1}, E_{12}, E_{22}, \ldots, E_{n2}, \ldots, E_{1n}, E_{2n}, \ldots, E_{nn}\},$$

where E_{ij} is the $n \times n$ matrix with the (i,j)-entry 1 and 0 elsewhere. It is easy to compute that

$$\mathcal{L}(E_{ij}) = AE_{ij} = a_{1i}E_{1j} + a_{2i}E_{2j} + \cdots + a_{ni}E_{nj}.$$

Thus the matrix representation of $\mathcal{L}$ under the basis is

$$\begin{pmatrix} A & & & 0 \\ & A & & \\ & & \ddots & \\ 0 & & & A \end{pmatrix}.$$

The desired result follows immediately.

3.111 (a) and (b) can be directly verified.

(c) $\mathcal{B}$ is one-to-one, but not onto.

(d) For every $f(x)$, $(\mathcal{AB} - \mathcal{BA})(f) = f$.

(e) By induction on k.

3.112 Obviously, $\operatorname{Ker}\mathcal{A} \subseteq \operatorname{Ker}\mathcal{A}^2 \subseteq \cdots \subseteq \operatorname{Ker}\mathcal{A}^k \subseteq \operatorname{Ker}\mathcal{A}^{k+1} \subseteq \cdots$. This inclusion chain stops because the dimensions are finite numbers. Thus $\operatorname{Ker}\mathcal{A}^m = \operatorname{Ker}\mathcal{A}^{m+k}$ for some m and for all positive integers k. We claim that $V = \operatorname{Ker}\mathcal{A}^m \oplus \operatorname{Im}\mathcal{A}^m$. All we need to show is $\operatorname{Ker}\mathcal{A}^m \cap \operatorname{Im}\mathcal{A}^m = \{0\}$. Let $u \in \operatorname{Ker}\mathcal{A}^m \cap \operatorname{Im}\mathcal{A}^m$. Then $\mathcal{A}^m(u) = 0$ and $u = \mathcal{A}^m(v)$ for some v. So $\mathcal{A}^m(u) = \mathcal{A}^{2m}(v) = 0$, which implies $v \in \operatorname{Ker}\mathcal{A}^{2m} = \operatorname{Ker}\mathcal{A}^m$. Thus $\mathcal{A}^m(v) = 0$; that is, $u = \mathcal{A}^m(v) = 0$.

Hints and Answers for Chapter 4

4.1 It is sufficient to show that (c)⇒(a) and (g)⇒(a).

(c)⇒(a): It is easy to see that $a_{ss} \in \mathbb{R}$, $s = 1, 2, \ldots, n$, by taking x to be the column vector with the s-th component 1, and 0 elsewhere.

Now take x to be the column vector with the s-th component 1, the t-th component c, and 0 elsewhere, where $s \neq t$ and c is an arbitrary complex number. Then

$$x^* Ax = a_{ss} + a_{tt}|c|^2 + a_{ts}\bar{c} + a_{st}c \in \mathbb{R}.$$

Putting $c = 1$ and i gives $a_{st} = \overline{a_{ts}}$, or $A^* = A$.

(g)⇒(a): Let $A = U^*TU$, where U is a unitary matrix and $T = (t_{st})$ is an upper-triangular matrix. Let the eigenvalues of A; that is, the diagonal entries of T, be $\lambda_1, \lambda_2, \ldots, \lambda_n$. Then

$$\operatorname{tr} A^2 = \operatorname{tr}(AA^*) \ \Rightarrow\ \operatorname{tr} T^2 = \operatorname{tr}(TT^*),$$

which is

$$\sum_{s=1}^{n} t_{ss}^2 = \sum_{s=1}^{n} |t_{ss}|^2 + \sum_{s<t} |t_{st}|^2.$$

It is immediate that $t_{st} = 0$ for every pair of s and t, $s < t$, and t_{ss} is real for each s. Therefore, T is real diagonal and A is Hermitian.

No. $A = \left(\begin{smallmatrix} 0 & 1 \\ 0 & 0 \end{smallmatrix}\right)$ is not Hermitian, but $A^*(A^2) = A^*(A^*A)$.

4.2
(a) True.
(b) False in general. True when c is real.
(c) False.
(d) True.
(e) False. Take $A = \left(\begin{smallmatrix} 1 & 1 \\ 1 & 1 \end{smallmatrix}\right)$, $B = \begin{pmatrix} 1 & -1 \\ -1 & 1 \end{pmatrix}$.
(f) True. $(BA)^* = A^*B^* = AB = 0$.
(g) True. Consider the diagonal case.
(h) False. Take $A = \left(\begin{smallmatrix} 0 & 1 \\ 1 & 0 \end{smallmatrix}\right)$.
(i) True. The eigenvalues of A are all 1.

(j) True.

(k) True.

(l) True.

(m) False. Take $A = \begin{pmatrix} 0 & 1 \\ 1 & 0 \end{pmatrix}$, $B = \begin{pmatrix} 1 & 0 \\ 0 & -1 \end{pmatrix}$.

(n) True.

(o) True. Consider the case where A is real diagonal.

(p) True, for BAB is Hermitian.

4.3 If λ is an eigenvalue of A, then λ^2 is an eigenvalue of A^2. Thus the eigenvalues of A^2 are $\lambda_1^2, \lambda_2^2, \dots, \lambda_n^2$. So $\operatorname{tr} A^2 = \lambda_1^2 + \lambda_2^2 + \cdots + \lambda_n^2$. However, $\operatorname{tr} A^2$ is the sum of the entries on the main diagonal of A^2. The desired identity follows. When A is Hermitian, $a_{ij} = \bar{a}_{ji}$.

4.4 By the spectral decomposition, let $A = U^* \operatorname{diag}(\lambda_1, \dots, \lambda_n) U$, where U is a unitary matrix. Then for $x \in \mathbb{C}^n$ with $\|x\| = 1$,

$$\begin{aligned} x^*Ax &= x^*U^* \operatorname{diag}(\lambda_1, \dots, \lambda_n)Ux \\ &= y^* \operatorname{diag}(\lambda_1, \dots, \lambda_n)y \\ &= \lambda_1|y_1|^2 + \cdots + \lambda_n|y_n|^2, \end{aligned}$$

where $y = (y_1, \dots, y_n)^t = Ux$ is also a unit vector. Thus

$$\min\{\lambda_1, \dots, \lambda_n\} \le x^*Ax \le \max\{\lambda_1, \dots, \lambda_n\}.$$

If x_0 is an eigenvector belonging to $\lambda_{\min}$ and $\|x_0\| = 1$, then

$$x_0^*Ax_0 = \lambda_{\min}.$$

It follows that

$$\lambda_{\min} = \min_{\|x\|=1} x^*Ax.$$

Similarly,

$$\lambda_{\max} = \max_{\|x\|=1} x^*Ax.$$

To see the inequalities for $A + B$,

$$\begin{aligned} \lambda_{\max}(A+B) &= \max_{\|x\|=1} x^*(A+B)x \\ &= \max_{\|x\|=1} (x^*Ax + x^*Bx) \\ &\le \max_{\|x\|=1} (x^*Ax) + \max_{\|x\|=1} (x^*Bx) \\ &= \lambda_{\max}(A) + \lambda_{\max}(B). \end{aligned}$$

The other inequality is similarly obtained by noting that

$$\max_{\|x\|=1} (x^*Ax + x^*Bx) \geq \max_{\|x\|=1} (x^*Ax) + \min_{\|x\|=1} (x^*Bx).$$

4.5 (a) If $x^tAx = 0$ for every $x \in \mathbb{R}^n$, then the diagonal entries of A are all zero by taking x to be the column vector with the s-th component 1, 0 elsewhere, $s = 1, 2, \ldots, n$.

Now take x to be the column vector with the s-th and the t-th components 1 and 0 elsewhere. Then

$$x^tAx = 0 \Rightarrow a_{st} = -a_{ts}, \quad s \neq t,$$

that is, $A^t = -A$. Conversely, if $A^t = -A$, then

$$x^tAx = (x^tAx)^t = x^tA^tx = -(x^tAx)$$

and $x^tAx = 0$ for every $x \in \mathbb{R}^n$.

(b) To show $a_{st} = 0$, take x and y to be the vectors whose s-th and t-th components are 1, respectively, and 0 elsewhere.

(c) It is easy to see that the diagonal entries of A are all equal to zero. Take x to be the column vector with the s-th component 1 and the t-th component c, then

$$x^*Ax = 0 \Rightarrow a_{ts}\bar{c} + a_{st}c = 0, \quad \text{for every } c \in \mathbb{C},$$

thus $a_{ts} = a_{st} = 0$.

(d) Let $x^*Ax = c$ be a constant. Then

$$x^*(A - cI)x = 0.$$

It follows from (c) that $A = cI$.

A is not necessarily equal to B even though $x^*Ax = x^*Bx$ for all $x \in \mathbb{R}^n$. Take $A = \begin{pmatrix} 0 & 1 \\ -1 & 0 \end{pmatrix}$ and $B = 0$.

4.6 Use the decomposition $A = U^* \operatorname{diag}(\lambda_1, \lambda_2, \ldots, \lambda_n)U$.

4.7 Since A is Hermitian, all eigenvalues of A are real. If $|A| < 0$, then at least one eigenvalue is negative. Denote it by λ. Then $Ax = \lambda x$ for some nonzero x, where $\lambda < 0$. Thus $x^*Ax = \lambda x^*x < 0$.

4.8 $(A + B)^* = A^* + B^* = A + B$. So $A + B$ is Hermitian. If $AB = BA$, then $(AB)^* = B^*A^* = BA = AB$; this says AB is Hermitian. Conversely, if AB is Hermitian, then $AB = (AB)^* = B^*A^* = BA$.

4.9 Since A and B are Hermitian matrices, AB is Hermitian if and only if $AB = BA$. Thus A and B are diagonalizable through the same unitary matrix; that is, U^*AU and U^*BU are diagonal for some unitary matrix U. Therefore the eigenvalues of AB are those in the form $\lambda = ab$, where a is an eigenvalue of A and b is an eigenvalue of B.

4.10 (a) There exists an orthogonal matrix P such that

$$A = P^{-1} \operatorname{diag}(6, 0, 0)P.$$

Take

$$X = P^{-1} \operatorname{diag}(6^{\frac{1}{3}}, 0, 0)P = P^t \operatorname{diag}(6^{\frac{1}{3}}, 0, 0)P.$$

(b) Suppose $X^3 = B$. Then the eigenvalues of X are all 0, this implies that $X^3 = 0$.

(c) $B = \begin{pmatrix} 0 & 0 & 1 \\ 0 & 0 & 0 \\ 0 & 1 & 0 \end{pmatrix}^2.$

(d) As (b).

(e) Yes. $D = \begin{pmatrix} 0 & 0 & 1 & 0 \\ 0 & 0 & 0 & 1 \\ 0 & 1 & 0 & 0 \\ 0 & 0 & 0 & 0 \end{pmatrix}^2.$

(f) Let $A = P^{-1} \operatorname{diag}(\lambda_1, \ldots, \lambda_n)P$, where λ_i's are real numbers and P is an orthogonal matrix. Then a k-th root of A is

$$X = P^{-1} \operatorname{diag}(\lambda_1^{\frac{1}{k}}, \ldots, \lambda_n^{\frac{1}{k}})P.$$

(g) If $X^2 = Y$, then $\lambda^2 I - X^2 = \lambda^2 I - Y$, which yields

$$(\lambda I - X)(\lambda I + X) = \lambda^2 I - Y.$$

Take determinants for both sides.

It is easy to check by computation that $\left(\begin{smallmatrix} -1 & -1 \\ 1 & 0 \end{smallmatrix}\right)^3 = I_2$.

4.11 Consider the case where A is real diagonal.

4.12 By direct computations.

4.13 Let U be a unitary matrix such that $A = U^* \operatorname{diag}(\lambda_1, \lambda_2, \lambda_3)U$. Then

$$tI - A = U^* \operatorname{diag}(t - \lambda_1,\ t - \lambda_2,\ t - \lambda_3)U$$

and for $t \neq \lambda_i,\ i = 1, 2, 3,$

$$\operatorname{adj}(tI - A) = |tI - A|(tI - A)^{-1}.$$

Suppose, without loss of generality, that A_1 is the 2×2 submatrix of A in the upper-left corner with eigenvalues a and b. Upon computation, the (3, 3)-entry of $(tI - A)^{-1}$ is

$$\frac{|u_{13}|^2}{t - \lambda_1} + \frac{|u_{23}|^2}{t - \lambda_2} + \frac{|u_{33}|^2}{t - \lambda_3}$$

and the (3, 3)-entry of $\operatorname{adj}(tI - A)$ is $|tI - A_1|$. So if $t \neq \lambda_i,\ i = 1, 2, 3,$

$$\frac{|tI - A_1|}{|tI - A|} = \frac{|u_{13}|^2}{t - \lambda_1} + \frac{|u_{23}|^2}{t - \lambda_2} + \frac{|u_{33}|^2}{t - \lambda_3}.$$

If a and b are roots of $|tI - A_1|$, it follows from the above identity that $a \in [\lambda_1, \lambda_2]$ and $b \in [\lambda_2, \lambda_3]$.

4.14 Suppose that H is an $m \times m$ principal submatrix of A. Let $x_m \in \mathbb{C}^n$ be a vector whose last $n - m$ components are 0. Then

$$\begin{aligned}
\lambda_{\max}(A) &= \max_{\|x\|=1,\ x \in \mathbb{C}^n} x^* A x \\
&\geq \max_{\|x_m\|=1} x_m^* A x_m \\
&= \max_{\|x\|=1,\ x \in \mathbb{C}^m} x^* H x \\
&= \lambda_{\max}(H).
\end{aligned}$$

A similar argument yields the other inequality.

4.15 (a) Take $A = \operatorname{diag}(1, -1)$; and I_2

(b) Assume that $A = U^* D U$, where D is real diagonal and U is unitary. Then $AB = U^* D U B$ is similar to $DUBU^*$ whose trace is real. In general,

$$\operatorname{tr}(AB)^k = \operatorname{tr}(AB \cdots AB) = \operatorname{tr}[(AB \cdots ABA)B]$$

is real since both $AB \cdots ABA$ and B are Hermitian.

(c) Assume, without loss of generality, that A is a real diagonal matrix with diagonal entries $a_1, a_2, \ldots, a_n$. Then

$$\begin{aligned} \operatorname{tr}(A^2B^2) - \operatorname{tr}(AB)^2 &= \sum_{i,j} a_i^2 |b_{ij}|^2 - \sum_{i,j} a_i a_j |b_{ij}|^2 \\ &= \sum_{i<j} (a_i - a_j)^2 |b_{ij}|^2 \geq 0. \end{aligned}$$

Equality holds if and only if $(a_i - a_j)|b_{ij}| = 0$ or all $i < j$, which is true if and only if $a_i b_{ij} = a_j b_{ij}$ for all i, j; that is, $AB = BA$.

(d) Assume that A is a real diagonal matrix with diagonal entries $a_1, a_2, \ldots, a_n$. It must be shown that

$$\Big(\sum_{i=1}^n a_i b_{ii}\Big)^2 \leq \Big(\sum_{i=1}^n a_i^2\Big)\Big(\sum_{i,\, j=1}^n |b_{ij}|^2\Big).$$

This is readily seen from the fact that

$$\Big(\sum_{i=1}^n a_i b_{ii}\Big)^2 \leq \Big(\sum_{i=1}^n a_i^2\Big)\Big(\sum_{i=1}^n b_{ii}^2\Big).$$

Both equalities hold if and only if the vectors $(a_1, \ldots, a_n)$ and $(b_{11}, \ldots, b_{nn})$ are linearly dependent and all $b_{ij} = 0$ when $i \neq j$.

The eigenvalues of AB are not real in general (but $\operatorname{tr}(AB)$ is). Take

$$A = \begin{pmatrix} 0 & 1 \\ 1 & 0 \end{pmatrix}, \quad B = \begin{pmatrix} 1 & 0 \\ 0 & -1 \end{pmatrix}.$$

4.16 By the spectral decomposition, write $A = UDU^*$, where U is unitary and $D = \operatorname{diag}(\lambda_1, \ldots, \lambda_r, 0, \ldots, 0)$, $\lambda_i \neq 0$, $i = 1, \ldots, r$. Let $\Lambda = \operatorname{diag}(\lambda_1, \ldots, \lambda_r)$ and denote the $n \times r$ matrix of the first r columns of U by U_r. Then

$$A = UDU^* = U_r \Lambda U_r^*.$$

Let $[A]_\alpha$ be an $r \times r$ principal submatrix of A lying in the intersections of the rows and columns $\alpha_1, \ldots, \alpha_r$. Denote by V_r the submatrix of U_r that consists of the rows of $\alpha_1, \ldots, \alpha_r$ of U_r. Then $[A]_\alpha = V_r \Lambda V_r^*$ and $\det[A]_\alpha = \det \Lambda |\det V_r|^2 = \lambda_1 \cdots \lambda_r |\det V_r|^2$. If $[A]_\alpha$ is nonsingular, then $\det[A]_\alpha$ has the same sign as $\lambda_1 \cdots \lambda_r$. It follows that all nonzero $r \times r$ principal minors of A have the same sign.

4.17 Denote the sum of the entries of row i of $A \in M_n(\mathbb{R})$ by r_i. Then

$$S(A) = r_1 + r_2 + \cdots + r_n.$$

Since $a_{ik} = a_{ki}$ and by interchanging the summands,

$$\begin{aligned} S(A^2) &= \sum_{i=1}^{n}\sum_{j=1}^{n}\sum_{k=1}^{n} a_{ik}a_{kj} \\ &= \sum_{k=1}^{n}\Big(\sum_{i=1}^{n}\sum_{j=1}^{n} a_{ki}a_{kj}\Big) \\ &= \sum_{k=1}^{n}\Big(\sum_{i=1}^{n} a_{ki}\Big)\Big(\sum_{j=1}^{n} a_{kj}\Big) \\ &= \sum_{k=1}^{n} r_k^2 \\ &\geq \frac{1}{n}\Big(\sum_{k=1}^{n} r_k\Big)^2 \quad \text{(the Cauchy-Schwarz inequality)} \\ &= \frac{1}{n}[S(A)]^2. \end{aligned}$$

4.18 We only show that (a)$\Rightarrow$(e), (f)$\Rightarrow$(b), (g)$\Rightarrow$(a), and (h)$\Leftrightarrow$(a).

(a)$\Rightarrow$(e): Let $A = U^* \operatorname{diag}(D, 0)U$ be the spectral decomposition of A, where D is $r \times r$, positive diagonal, nonsingular. Let U_1 be the first r rows of U. Then $A = U_1^* D^{1/2} \cdot D^{1/2} U_1 = T^*T$, where $T = D^{1/2}U_1$.

To see (f)$\Rightarrow$(b), let $A = U^* \operatorname{diag}(\lambda_1, \ldots, \lambda_n)U$, where λ_i's are (real) eigenvalues of A and U is unitary, and let

$$f(\lambda) = |\lambda I - A| = \lambda^n - \alpha_1\lambda^{n-1} + \alpha_2\lambda^{n-2} - \cdots + (-1)^n\alpha_n.$$

It can be shown that α_i is the sum of all $i \times i$ minors of A, $i = 1, 2, \ldots, n$. Thus if all the minors of A are nonnegative, then $f(\lambda)$ has no negative zeros.

For (g)$\Rightarrow$(a): Take $X = xx^*$, where x is any column vector.

For (h)$\Leftrightarrow$(a): If $x^*Ax \geq 0$ for all vector x, then $y^*(X^*AX)y = (Xy)^*A(Xy) \geq 0$ for all y; this says $X^*AX \geq 0$. Conversely, if for some (fixed) positive integer m, $X^*AX \geq 0$ for all $n \times m$ matrices X, then for any n-column vector, we take X to be the $n \times m$ matrix with first column x and everything else 0. It follows that $x^*Ax \geq 0$.

4.19 $(A+B)^2 = A^2 + AB + BA + B^2 \geq 0$. But $A^2 + AB + BA$ and $AB + BA$ are not necessarily positive semidefinite. Take

$$A = \begin{pmatrix} 1 & 1 \\ 1 & 1 \end{pmatrix} \geq 0, \qquad B = \begin{pmatrix} 2 & 1 \\ 1 & 1 \end{pmatrix} \geq 0.$$

Then

$$A^2 + AB + BA = \begin{pmatrix} 8 & 7 \\ 7 & 6 \end{pmatrix} \not\geq 0, \quad AB + BA = \begin{pmatrix} 6 & 5 \\ 5 & 4 \end{pmatrix} \not\geq 0.$$

If A and $AB + BA$ are positive definite, we show that B is positive definite. Let $C = AB + BA$. Multiply C by $A^{-\frac{1}{2}}$ from both sides:

$$0 < A^{-\frac{1}{2}} C A^{-\frac{1}{2}} = A^{\frac{1}{2}} B A^{-\frac{1}{2}} + A^{-\frac{1}{2}} B A^{\frac{1}{2}} = D + D^*,$$

where $D = A^{\frac{1}{2}} B A^{-\frac{1}{2}}$. It is sufficient to show that D is nonsingular. Suppose $Dx = 0$ for some nonzero column vector x. Then $x^* D^* = 0$ and $x^*(D + D^*)x = x^* D x + x^* D^* x = 0$. This contradicts $D + D^* > 0$.

4.20 Note that $(A \pm B)^*(A \pm B) \geq 0$. Expanding this yields the inequality.

4.21 Note that $0 \leq A \leq I \Rightarrow 0 \leq A^2 \leq A$. It follows that, by expanding,

$$0 \leq (A + B - 1/2)^2 \leq AB + BA + \frac{1}{4}.$$

4.22 $-2 < \lambda < 1$, $-\sqrt{2} - 1 < \mu < \sqrt{2} - 1$.

4.23 (a) is immediate from (b). So it is sufficient to show (b). Let $Ax = \lambda x$, where $x = (x_1, x_2, \ldots, x_n)^t \neq 0$. Since A is Hermitian, λ is real. Choose an i so that $|x_i| = \max_j |x_j|$. Then $x_i \neq 0$. From $\sum_{j=1}^n a_{ij} x_j = \lambda x_i$, we have $(\lambda - 1)x_i = \sum_{j=1, j\neq i}^n a_{ij} x_j$. By taking absolute values, we get $|\lambda - 1|\,|x_i| \leq |x_i|$. It follows that $|\lambda - 1| \leq 1$; that is, $0 \leq \lambda \leq 2$. So A is positive semidefinite. (c) follows from an application of the Hadamard inequality to $|A|$, which is the product of all eigenvalues of A. (See Problem 4.56.)

4.24 $\begin{pmatrix} 1 & 1 \\ 0 & 1 \end{pmatrix}$.

4.25 Yes, when $x \in \mathbb{R}^n$. For instance, $A = \begin{pmatrix} 0 & 1 \\ -1 & 0 \end{pmatrix}$. No, when $x \in \mathbb{C}^n$. In this case, A must be Hermitian.

4.26 (a) Consider each diagonal entry as a minor, or take x to be the column vector whose i-th component is 1, and 0 elsewhere. Then $a_{ii} = x^* A x$. The second part follows from (b).

(b) Consider the 2×2 minors.

(c) Write $A = U^* \operatorname{diag}(\lambda_1, \dots, \lambda_n)U$.

(d) Assume that B is the principal submatrix of A in the upper-left corner. If $|B| = 0$, then $Bv = 0$ for some $v \neq 0$. Set $x = (v^t, 0, \dots, 0)^t \in \mathbb{C}^n$. Then $x \neq 0$, $Ax = 0$. So A is singular.

(e) A is unitarily diagonalizable. Let $A = U^*DU$, where $D = \operatorname{diag}(\lambda_1, \lambda_2, \dots, \lambda_n)$. Split each $\lambda_i \neq 0$ as $\frac{1}{\sqrt{\lambda_i}} 1 \frac{1}{\sqrt{\lambda_i}}$. P cannot be unitary in general unless the eigenvalues of A are 1 or 0.

(f) $x^*Ax \geq 0 \Rightarrow (x^*Ax)^t \geq 0$. So $x^t A^t \bar{x} \geq 0$ or $y^* A^t y \geq 0$ for all $y = \bar{x}$. Thus $A^t \geq 0$. Likewise $\bar{A} \geq 0$.

4.27 $(Ax)^*(Ax) = (x^*A^*)(Ax) = 0$ if and only if $Ax = 0$. If $A \geq 0$, then $\operatorname{tr} A = 0$ if and only if $A = 0$.

4.28 Since the identity holds for every x if and only if it holds for Ux with y replaced by Uy, we may assume that A is a diagonal matrix.

Let $A = \operatorname{diag}(\lambda_1, \lambda_2, \dots, \lambda_n)$. Then for any $x \in \mathbb{C}^n$,

$$x^*Ax = \sum_{k=1}^{n} \lambda_k |x_k|^2.$$

Notice that

$$x_k \overline{y_k} + y_k \overline{x_k} - \lambda_k |y_k|^2 = -\lambda_k (y_k - \lambda_k^{-1} x_k)\overline{(y_k - \lambda_k^{-1} x_k)} + \lambda_k^{-1} |x_k|^2.$$

It follows, by taking sum and maximizing both sides, that

$$\max_y (x^*y + y^*x - y^*Ay) = \sum_{k=1}^{n} \lambda_k^{-1} |x_k|^2 = x^* A^{-1} x,$$

and the maximum is attained at $y = (y_1, y_2, \dots, y_n)$, where $y_k = \lambda_k^{-1} x_k$, $k = 1, 2, \dots, n$.

4.29 It is sufficient to show that $\operatorname{Im}(AB) \cap \operatorname{Ker}(AB) = \{0\}$. Let y be in the intersection and write $y = (AB)x$ for some x. Since $y \operatorname{Ker}(AB)$,

$(AB)y = (AB)^2x = 0$. We claim $(AB)x = 0$ as follows:

$$\begin{aligned}(AB)^2x = 0 \;&\Rightarrow\; (ABAB)x = 0\\ &\Rightarrow\; (x^*B)(ABAB)x = 0\\ &\Rightarrow\; (x^*BAB^{1/2})(B^{1/2}ABx) = 0\\ &\Rightarrow\; (B^{1/2}AB)x = 0\\ &\Rightarrow\; B^{1/2}(B^{1/2}AB)x = (BAB)x = 0\\ &\Rightarrow\; (x^*BA^{1/2})(A^{1/2}Bx) = 0\\ &\Rightarrow\; (A^{1/2}B)x = 0\\ &\Rightarrow\; (AB)x = 0.\end{aligned}$$

4.30 (a) Let $A = U^* \operatorname{diag}(\lambda_1, \ldots, \lambda_n)U$, where U is unitary. Take

$$B = U^* \operatorname{diag}(\lambda_1^{\frac{1}{2}}, \ldots, \lambda_n^{\frac{1}{2}})U.$$

To show the uniqueness, suppose $C \geq 0$ and $C^2 = B^2 = A$. Let

$$C = V \operatorname{diag}(\lambda_1^{\frac{1}{2}}, \ldots, \lambda_n^{\frac{1}{2}})V^* \text{ and } T = UV.$$

Then

$$T \operatorname{diag}(\lambda_1, \ldots, \lambda_n) = \operatorname{diag}(\lambda_1, \ldots, \lambda_n)T.$$

A direct computation gives $B = C$.

(b) Any normal matrix has a square root. It is neither unique nor positive semidefinite in general.

(c) The square roots are, respectively,

$$\begin{pmatrix} \sqrt{2} & 0 \\ 0 & 0 \end{pmatrix}, \quad \frac{1}{\sqrt{2}}\begin{pmatrix} 1 & 1 \\ 1 & 1 \end{pmatrix}, \quad \frac{1}{2}\begin{pmatrix} 3 & -1 \\ -1 & 3 \end{pmatrix}.$$

4.31 $A^2C = A(AC) = A(CA) = (AC)A = (CA)A = CA^2$. For the square root, let $A = U^*DU$, where U is unitary, $D = \operatorname{diag}(d_1, d_2, \ldots, d_n)$, where each $d_i \geq 0$. Let $W = UCU^*$. Then $AC = CA$ gives $DW = WD$. So $d_i w_{ij} = w_{ij} d_j$. This implies $\sqrt{d_i}w_{ij} = w_{ij}\sqrt{d_j}$; that is, $D^{\frac{1}{2}}W = WD^{\frac{1}{2}}$. This immediately yields $A^{\frac{1}{2}}C = CA^{\frac{1}{2}}$.

4.32 (a) If $A = A^*$, then $A^2 = A^*A \geq 0$.

(b) If $A^* = -A$, then $-A^2 = (-A)A = A^*A \geq 0$.

(c) Obvious.

None of the converses is true. For (a) and (b), take

$$\begin{pmatrix} 0 & 1 & 0 \\ 0 & 0 & 0 \\ 0 & 0 & -1 \end{pmatrix} \quad \text{or} \quad i \begin{pmatrix} 1 & 2 & -1 \\ 0 & -1 & 1 \\ 0 & 0 & 1 \end{pmatrix}.$$

The square of the latter matrix is $-I_3$. For (c), take

$$\begin{pmatrix} 1 & 0 & -1 \\ 2 & -1 & 1 \\ 0 & 0 & 1 \end{pmatrix},$$

whose diagonal entries are the same as the eigenvalues.

4.33 It must be shown that

$$(tA + \tilde{t}B)^*(tA + \tilde{t}B) \leq tA^*A + \tilde{t}B^*B$$

or

$$t^2A^*A + t\tilde{t}(B^*A + A^*B) + \tilde{t}^2B^*B \leq tA^*A + \tilde{t}B^*B,$$

which is

$$0 \leq t\tilde{t}(A^*A + B^*B - B^*A - A^*B) = t\tilde{t}(A^* - B^*)(A - B).$$

This is always true.

4.34 $tA^2 + (1-t)B^2 - (tA + (1-t)B)^2 = t(1-t)(A-B)^2 \geq 0$. This yields the first inequality. Note that $(a-b)I \leq A - B \leq (b-a)I$ gives $(A-B)^2 \leq (b-a)^2 I$ and $t(1-t) \leq \frac{1}{4}$. The second inequality follows.

4.35 (a) Let T be an invertible matrix such that $A = T^*T$. Since

$$(T^{-1})^*BT^{-1} \geq 0,$$

we have a unitary matrix U such that $(T^{-1})^*BT^{-1} = UDU^*$, where D is a diagonal matrix with nonnegative entries. Put $P = T^{-1}U$. Then $P^*AP = I, \quad P^*BP = D$.

(b) Use induction on n. Let S be an invertible matrix such that

$$S^*AS = \begin{pmatrix} 0 & 0 \\ 0 & I_r \end{pmatrix}, \quad \text{where } r = r(A) < n.$$

Write

$$S^*BS = \begin{pmatrix} b_{11} & \alpha \\ \alpha^* & B_1 \end{pmatrix}.$$

If $b_{11} = 0$, then $\alpha = 0$. The conclusion follows by induction on B_1 and A_1, which is obtained by deleting the first row and first column of S^*AS. If $b_{11} \neq 0$, let

$$T = \begin{pmatrix} 1 & -b_{11}^{-1}\alpha \\ 0 & I_{n-1} \end{pmatrix}.$$

Then

$$T^*S^*AST = S^*AS$$

and

$$T^*S^*BST = \begin{pmatrix} b_{11} & 0 \\ 0 & B_1 - \alpha^* b_{11}^{-1}\alpha \end{pmatrix} \geq 0.$$

Notice that $B_1 - \alpha^* b_{11}^{-1}\alpha \geq 0$. By induction, there exists an invertible $(n-1) \times (n-1)$ matrix P_1 such that $P_1^* A_1 P_1$ and $P_1^*(B_1 - \alpha^* b_{11}^{-1}\alpha)P_1$ are both diagonal. Now set

$$P = ST \begin{pmatrix} 1 & 0 \\ 0 & P_1 \end{pmatrix}.$$

The desired conclusion follows immediately.

The $B \geq 0$ cannot be changed by a Hermitian B. Take

$$A = \begin{pmatrix} 1 & 0 \\ 0 & 0 \end{pmatrix} \text{ and } B = \begin{pmatrix} 0 & 1 \\ 1 & 0 \end{pmatrix}.$$

(c) It is immediate from (a).

(d) By (b).

(e) It follows from (b) and the Hölder inequality (which can be proved by induction): for nonnegative numbers a's and b's,

$$(a_1 \cdots a_n)^{\frac{1}{n}} + (b_1 \cdots b_n)^{\frac{1}{n}} \leq [(a_1 + b_1) \cdots (a_n + b_n)]^{\frac{1}{n}}.$$

(f) Use (b) and the fact that $a^t b^{\tilde{t}} \leq ta + \tilde{t}b$ for $a, b \geq 0$ and $t \in [0, 1]$. For the particular case, take $t = \frac{1}{2}$. Note that $2^k \leq 2^n$.

(g) Since $\sqrt{ab} \leq \frac{1}{2}(a + b)$ when a, $b \geq 0$ and by (d).

4.36 $A = (A - B) + B$. By Problem 4.35, $|A| \geq |B|$. Since B is positive definite, there exists a nonsingular matrix P such that $P^*BP = I$. Let $C = P^*(A-B)P$. Then $C \geq 0$. Since $A - \lambda B = (A-B) - (\lambda - 1)B$, we have $|P^*|\,|A - \lambda B|\,|P| = |C - (\lambda - 1)I|$. Thus $\lambda - 1 \geq 0$ and $\lambda \geq 1$.

4.37 (a) $x^*(C^*AC - C^*BC)x = (Cx)^*(A - B)(Cx) \geq 0$.

(b) $(A - B) + (C - D) \geq 0$.

(c) $A - B \geq 0 \Rightarrow \operatorname{tr}(A - B) \geq 0 \Rightarrow \operatorname{tr} A \geq \operatorname{tr} B$.

(d) $A - B \geq 0 \Rightarrow x^*(A - B)x \geq 0$, or $x^*Ax \geq x^*Bx$. Thus

$$\lambda_{\max}(A) = \max_{\|x\|=1} x^*Ax \geq \max_{\|x\|=1} x^*Bx = \lambda_{\max}(B).$$

(e) Note that $|A| = |(A - B) + B| \geq |B|$. An alternative proof: If $|B| = 0$, there is nothing to show. Assume that $|B| \neq 0$. Then

$$A \geq B \quad \Rightarrow \quad B^{-\frac{1}{2}}AB^{-\frac{1}{2}} \geq I$$

and

$$|B^{-\frac{1}{2}}AB^{-\frac{1}{2}}| \geq 1 \quad \Rightarrow \quad |A| \geq |B|.$$

(f) Use Problem 4.35.

(g) This can be proved in different ways. A directly proof: If $B = I$,

$$A \geq I \quad \Rightarrow \quad I \geq A^{-\frac{1}{2}}IA^{-\frac{1}{2}},$$

namely, $I \geq A^{-1}$. In general,

$$A \geq B \quad \Rightarrow \quad B^{-\frac{1}{2}}AB^{-\frac{1}{2}} \geq I.$$

Thus $I \geq B^{\frac{1}{2}}A^{-1}B^{\frac{1}{2}}$ and $B^{-1} \geq A^{-1}$.

(h) First note that $A^{\frac{1}{2}} - B^{\frac{1}{2}}$ is Hermitian. It must be shown that the eigenvalues of $A^{\frac{1}{2}} - B^{\frac{1}{2}}$ are nonnegative. Let

$$(A^{\frac{1}{2}} - B^{\frac{1}{2}})x = \lambda x, \quad x \neq 0.$$

Then

$$B^{\frac{1}{2}}x = A^{\frac{1}{2}}x - \lambda x.$$

Notice that (the Cauchy-Schwarz inequality)

$$|x^*y| \leq (x^*x)^{\frac{1}{2}}\,(y^*y)^{\frac{1}{2}}, \quad x, y \in \mathbb{C}^n.$$

Since $A \geq B$, we have $(x^*Ax)^{\frac{1}{2}} \geq (x^*Bx)^{\frac{1}{2}}$ and

$$\begin{aligned} x^*Ax &= (x^*Ax)^{\frac{1}{2}}(x^*Ax)^{\frac{1}{2}} \\ &\geq (x^*Ax)^{\frac{1}{2}}(x^*Bx)^{\frac{1}{2}} \\ &\geq |(x^*A^{\frac{1}{2}})(B^{\frac{1}{2}}x)| \\ &= |x^*A^{\frac{1}{2}}(A^{\frac{1}{2}}x - \lambda x)| \\ &= |x^*Ax - \lambda x^*A^{\frac{1}{2}}x|. \end{aligned}$$

Thus either $\lambda = 0$ or $\lambda x^*A^{\frac{1}{2}}x = 2x^*Ax \geq 0$, so $\lambda \geq 0$.
It is not true that $A^2 \geq B^2$ in general. Take

$$A = \begin{pmatrix} 2 & 1 \\ 1 & 1 \end{pmatrix}, \quad B = \begin{pmatrix} 1 & 1 \\ 1 & 1 \end{pmatrix}.$$

4.38 (a) $AB = A^{\frac{1}{2}}A^{\frac{1}{2}}B$ and $A^{\frac{1}{2}}(AB)BA^{\frac{1}{2}}$ have the same eigenvalues, while the latter one is Hermitian, thus all eigenvalues are real.

(b) $A + B \geq 0 \Leftrightarrow I + A^{-\frac{1}{2}}BA^{-\frac{1}{2}} \geq 0 \Leftrightarrow$ all the eigenvalues of $A^{-\frac{1}{2}}BA^{-\frac{1}{2}}$ are greater than or equal to -1. Now it is sufficient to notice that $A^{-\frac{1}{2}}BA^{-\frac{1}{2}}$ and $A^{-1}B$ have the same eigenvalues.

(c) $r(AB) = r(B) = r(A^{1/2}BA^{1/2})$. The latter equals the number of nonzero eigenvalues of $A^{1/2}BA^{1/2}$, as it is Hermitian. Note that $A^{1/2}BA^{1/2}$ and AB have the same number of nonzero eigenvalues. If $A \geq 0$, then it is not true. Take $A = \left(\begin{smallmatrix} 1 & 0 \\ 0 & 0 \end{smallmatrix}\right)$ and $B = \left(\begin{smallmatrix} 0 & 1 \\ 1 & 0 \end{smallmatrix}\right)$. Then $r(AB) = 1$, while AB has no nonzero eigenvalues.

4.39 (a) $A = \begin{pmatrix} 1 & 0 \\ 0 & -2 \end{pmatrix}$, $B = \begin{pmatrix} 1 & 1 \\ 1 & -1 \end{pmatrix}$. The eigenvalues of AB are $\frac{1}{2}(3 \pm \sqrt{7}i)$.

(b) Let $A \geq 0$. The AB has the same eigenvalues as the Hermitian matrix $A^{\frac{1}{2}}BA^{\frac{1}{2}}$. The eigenvalues of the latter are all real.

(c) Let $A > 0$. Then AB is similar to $A^{-\frac{1}{2}}(AB)A^{\frac{1}{2}} = A^{\frac{1}{2}}BA^{\frac{1}{2}}$, which is Hermitian, and of course diagonalizable.

(d) $A = \left(\begin{smallmatrix} 1 & 1 \\ 1 & 1 \end{smallmatrix}\right)$, $B = \begin{pmatrix} 1 & 0 \\ 0 & -1 \end{pmatrix}$. The eigenvalues of AB are $0, 0$. AB cannot be diagonalizable, since $AB \neq 0$.

4.40 Let $c = x^*Ax$, where $\|x\| = 1$. By the Cauchy-Schwarz inequality

$$x^*\Big(\frac{A + A^*}{2}\Big)x = \frac{c + \bar{c}}{2} \leq |c| = |\bar{c}c|^{\frac{1}{2}} \leq (x^*A^*Ax)^{\frac{1}{2}}.$$

Thus

$$\begin{aligned}\lambda_{\max}\Big(\frac{A+A^*}{2}\Big) &= \max_{x^*x=1} x^*\Big(\frac{A+A^*}{2}\Big)x\\ &\le \max_{x^*x=1}(x^*A^*Ax)^{\frac{1}{2}}\\ &\le \Big(\max_{x^*x=1} x^*A^*Ax\Big)^{\frac{1}{2}}\\ &= \sigma_{\max}(A).\end{aligned}$$

For the trace inequality, noting that $A - A^*$ is skew-Hermitian, we have $\operatorname{tr}(A - A^*)^2 \le 0$, which implies, by expanding and taking trace,

$$\operatorname{tr} A^2 + \operatorname{tr}(A^*)^2 \le 2 \operatorname{tr} A^*A,$$

equivalently,

$$\operatorname{tr}\Big(\frac{A+A^*}{2}\Big)^2 \le \operatorname{tr}(A^*A).$$

4.41 (a) $(A^{\frac{1}{2}})^* = A^{\frac{1}{2}}$.

(b) $AB = A^{\frac{1}{2}}(A^{\frac{1}{2}}B)$ has the same eigenvalues as $A^{\frac{1}{2}}BA^{\frac{1}{2}} \ge 0$.

(c) AB is not positive semidefinite in general. Take

$$A = \begin{pmatrix} 1 & 0 \\ 0 & 0 \end{pmatrix}, \quad B = \begin{pmatrix} 1 & 1 \\ 1 & 1 \end{pmatrix}.$$

(d) If A and B commute, then AB is Hermitian, since

$$(AB)^* = B^*A^* = BA = AB.$$

As AB and $A^{\frac{1}{2}}BA^{\frac{1}{2}}$ have the same eigenvalues, $AB \ge 0$ by (a). Conversely, if $AB \ge 0$, then AB is Hermitian and it follows that

$$AB = (AB)^* = B^*A^* = BA.$$

(e) Use $\operatorname{tr}(XY) = \operatorname{tr}(YX)$.

(f) Let $\lambda_1(X), \lambda_2(X), \ldots, \lambda_n(X)$ denote the eigenvalues of X. Since $AB^2A = (AB)(BA)$ and $BA^2B = (BA)(AB)$ have the same eigenvalues,

$$\begin{aligned}\operatorname{tr}(AB^2A)^{\frac{1}{2}} &= \sum_{i=1}^{n} \lambda_i[(AB^2A)^{\frac{1}{2}}]\\ &= \sum_{i=1}^{n} \sqrt{\lambda_i(AB^2A)}\\ &= \sum_{i=1}^{n} \sqrt{\lambda_i(BA^2B)}\\ &= \operatorname{tr}(BA^2B)^{\frac{1}{2}}.\end{aligned}$$

(g) It may be assumed that $A = \operatorname{diag}(\lambda_1, \ldots, \lambda_n)$. Suppose that $b_{11}, \ldots, b_{nn}$ are the main diagonal entries of B. Then

$$\begin{aligned}\operatorname{tr}(AB) &= \lambda_1 b_{11} + \cdots + \lambda_n b_{nn}\\ &\leq (\lambda_1 + \cdots + \lambda_n)(b_{11} + \cdots + b_{nn})\\ &= \operatorname{tr} A \operatorname{tr} B.\end{aligned}$$

(h) Assume that $A = \operatorname{diag}(\lambda_1, \ldots, \lambda_n)$. Then

$$\begin{aligned}\operatorname{tr}(AB) &= \lambda_1 b_{11} + \cdots + \lambda_n b_{nn}\\ &\leq \lambda_{\max}(A)(b_{11} + \cdots + b_{nn})\\ &= \lambda_{\max}(A) \operatorname{tr} B.\end{aligned}$$

(i)

$$\begin{aligned}\operatorname{tr}(AB) &= \lambda_1 b_{11} + \cdots + \lambda_n b_{nn}\\ &= \frac{1}{4}[(2\lambda_1 b_{11} + \cdots + 2\lambda_n b_{nn})\\ &\quad +(2\lambda_1 b_{11} + \cdots + 2\lambda_n b_{nn})]\\ &\leq \frac{1}{4}[(\lambda_1^2 + b_{11}^2 + \cdots + \lambda_n^2 + b_{nn}^2)\\ &\quad +(2\lambda_1 b_{11} + \cdots + 2\lambda_n b_{nn})]\\ &\leq \frac{1}{4}(\lambda_1 + \cdots + \lambda_n + b_{11} + \cdots + b_{nn})^2\\ &= \frac{1}{4}(\operatorname{tr} A + \operatorname{tr} B)^2.\end{aligned}$$

(j) Note that $A^2 + B^2 - AB - BA = (A - B)^2 \geq 0$. Take trace.

No. Take $A = \begin{pmatrix} \frac{3}{2} & 0 \\ 0 & \frac{3}{2} \end{pmatrix}$, $B = \begin{pmatrix} 3 & 0 \\ 0 & 0 \end{pmatrix}$.

4.42 (a) $(AB+BA)^* = B^*A^* + A^*B^* = BA + AB = AB + BA$.

(b) No, in general. Take $A = \begin{pmatrix} 2 & 0 \\ 0 & 1 \end{pmatrix}$, $B = \begin{pmatrix} 1 & 1 \\ 1 & 1 \end{pmatrix}$.

(c) No.

(d) Yes.

(e) Yes. $A^2 - AB - BA + B^2 = (A-B)^2 \geq 0$.

(f) Note that $\operatorname{tr}(XY) \geq 0$ when $X, Y \geq 0$. It follows that

$$\begin{aligned} \operatorname{tr}(CD) - \operatorname{tr}(AB) &= \operatorname{tr}(CD - CB) + \operatorname{tr}(CB - AB) \\ &= \operatorname{tr}[C(D-B)] + \operatorname{tr}[(C-A)B] \geq 0. \end{aligned}$$

(g) Since $\lambda(XY) = \lambda(YX)$,

$$\begin{aligned} \lambda_{\max}(B)I - B \geq 0 &\Rightarrow A^{\frac{1}{2}}[\lambda_{\max}(B)I - B]A^{\frac{1}{2}} \geq 0 \\ &\Rightarrow \lambda_{\max}(B)A \geq A^{\frac{1}{2}}BA^{\frac{1}{2}} \\ &\Rightarrow \lambda_{\max}(A)\lambda_{\max}(B) \geq \\ &\qquad \lambda_{\max}(A^{\frac{1}{2}}BA^{\frac{1}{2}}) = \lambda_{\max}(AB). \end{aligned}$$

(h) Use the result that $\lambda_{\max}(A) = \max_{\|x\|=1} x^*Ax$.

(i) For three positive semidefinite matrices, there is no similar result. In fact, the eigenvalues of ABC can be imaginary numbers. For instance, the eigenvalues of ABC are 0 and $8+i$, where

$$A = \begin{pmatrix} 1 & 1 \\ 1 & 1 \end{pmatrix}, \quad B = \begin{pmatrix} 2 & 1 \\ 1 & 1 \end{pmatrix}, \quad C = \begin{pmatrix} 2 & i \\ -i & 1 \end{pmatrix}.$$

4.43 (a) Take $A = \begin{pmatrix} 1 & 0 \\ 5 & 1 \end{pmatrix}$ and $B = \begin{pmatrix} 1 & -5 \\ 0 & 1 \end{pmatrix}$.

(b) No, because $\operatorname{tr}(A+B) = \operatorname{tr} A + \operatorname{tr} B > 0$.

(c) Take $A = \begin{pmatrix} 5 & 2 \\ 2 & 1 \end{pmatrix}$, $B = \begin{pmatrix} 3 & -1 \\ -1 & 1 \end{pmatrix}$, $C = \begin{pmatrix} 1 & 0 \\ 0 & 30 \end{pmatrix}$. Then A, B, and C are positive definite, the eigenvalues of ABC are $-5, -12$.

(d) No. Note that $|ABC| > 0$.

4.44 Since A is positive semidefinite, let $A = U^*DU$, where U is unitary and $D = \operatorname{diag}(\lambda_1, \lambda_2, \ldots, \lambda_n)$, $\lambda_i \geq 0$. Then $A^2B = BA^2$ if and only if $U^*D^2UB = BU^*D^2U$ if and only if $D^2(UBU^*) = (UBU^*)D^2$. Let $C = UBU^*$. Then $D^2C = CD^2$. We show that $DC = CD$. $D^2C = CD^2 \Rightarrow \lambda_i^2 c_{ij} = c_{ij}\lambda_j^2$ for all i and j. If $c_{ij} \neq 0$, then $\lambda_i = \lambda_j$. Thus $c_{ij}\lambda_i = c_{ij}\lambda_j$ for all i, j and $DC = CD$. It follows that $D(UBU^*) = (UBU^*)D$ or $AB = BA$.

The conclusion is not true in general for Hermitian matrix A. Take $A = \operatorname{diag}(1, -1)$. Then every 2×2 matrix commutes with A^2.

4.45 We show that C commutes with $A+B$ first. For this, we show that C commutes with $(A+B)^2$ and then C commutes with $A+B$, as one may prove if X commutes with a positive semidefinite matrix Y, then X commutes with the square root of Y. Since C is Hermitian and commutes with AB, $(AB)C = C(AB)$ implies $C^*(AB)^* = (AB)^*C^*$; that is, $C(BA) = (BA)C$. In other words, C commutes with BA. Now compute $C(A+B)^2$ and $(A+B)^2C$. Since C commutes with $A-B$, we have $C(A-B)^2 = (A-B)^2C$. Along with $CAB = ABC$ and $CBA = BAC$, we get $C(A+B)^2 = (A+B)^2C$. Thus $C(A+B) = (A+B)C$. It follows that C commutes with $2A = (A+B)+(A-B)$.

4.46 Let $Bx = \lambda x$, $x \neq 0$. Pre-postmultiplying $A > B^*AB$ by x^* and x, we have $x^*Ax > |\lambda|^2 x^*Ax$. Thus $|\lambda| < 1$.

This does not hold for singular values. Take $A = \left(\begin{smallmatrix} 1 & 0 \\ 0 & 5 \end{smallmatrix}\right)$ and $B = \left(\begin{smallmatrix} 0 & 2 \\ 0 & 0 \end{smallmatrix}\right)$. Then $A - B^*AB = I_2 > 0$. But the largest singular value of B is 2.

4.47 As A is a principal submatrix of the block matrix, A^{-1} exists. Thus

$$\begin{pmatrix} I & 0 \\ -B^*A^{-1} & I \end{pmatrix} \begin{pmatrix} A & B \\ B^* & C \end{pmatrix} \begin{pmatrix} I & -A^{-1}B \\ 0 & I \end{pmatrix}$$

$$= \begin{pmatrix} A & 0 \\ 0 & C - B^*A^{-1}B \end{pmatrix} > 0.$$

4.48 By a simple computation

$$(I,\ I) \begin{pmatrix} A & B \\ B^* & C \end{pmatrix} \begin{pmatrix} I \\ I \end{pmatrix} = A + B^* + B + C \geq 0.$$

4.49 Take

$$A = \begin{pmatrix} 1 & 1 \\ 1 & 2 \end{pmatrix}, \quad B = \begin{pmatrix} 1 & 1 \\ 2 & 2 \end{pmatrix}, \quad C = \begin{pmatrix} 3 & 3 \\ 3 & 3 \end{pmatrix}.$$

Then

$$\begin{pmatrix} A & B \\ B^* & C \end{pmatrix} \geq 0, \qquad \begin{pmatrix} A & B^* \\ B & C \end{pmatrix} \not\geq 0.$$

4.50 It is sufficient to note that

$$P^* \begin{pmatrix} A & B \\ B & A \end{pmatrix} P = \begin{pmatrix} A-B & 0 \\ 0 & A+B \end{pmatrix}, \quad P = \frac{1}{\sqrt{2}} \begin{pmatrix} I & I \\ -I & I \end{pmatrix}.$$

4.51 $A+iB \geq 0$ if and only if $x^*(A+iB)x \geq 0$ for all complex column vectors x. Taking conjugate gives $y^*(A-iB)y \geq 0$, $y=\bar{x}$. So $A+iB \geq 0$ if and only if $A-iB \geq 0$. Since $A+iB \geq 0$, $A=A^t$ and $B^t=-B$. So the partitioned matrix $M=\begin{pmatrix} A & -B \\ B & A \end{pmatrix}$ is real symmetric. Conversely, if M is real symmetric, then $A+iB$ is Hermitian. We show that all the eigenvalues of M are nonnegative if $A+iB \geq 0$ (then so $A-iB \geq 0$). This is seen through similarity by observing

$$M = S^{-1}\begin{pmatrix} A+iB & 0 \\ 0 & A-iB \end{pmatrix} S, \quad S=\frac{1}{\sqrt{2}}\begin{pmatrix} I & iI \\ -iI & -I \end{pmatrix}.$$

4.52 $\lambda_i+\frac{1}{\lambda_i}$, $i=1,2,\ldots,n$, plus n zeros.

4.53 (a) By computation,

$$|\lambda I - M| = \begin{vmatrix} \lambda I & -A^* \\ -A & \lambda I \end{vmatrix} = |\lambda^2 I - AA^*|.$$

Putting $\lambda = 0$ in the above identity gives

$$(-1)^{2n}|M| = (-1)^n |AA^*|$$

or

$$|M| = (-1)^n |\det A|^2.$$

(b) The eigenvalues of M are

$$\sigma_1(A),\ldots,\sigma_n(A),\ -\sigma_n(A),\ldots,-\sigma_1(A),$$

where $\sigma_i(A)$'s are the singular values of A.

(c) If $A \neq 0$, then M has negative eigenvalues.

(d) The eigenvalues of N are $1 \pm \sigma_i(A)$, $i=1,2,\ldots,n$.

4.54 First inequality: $\sigma^2_{\max}(AB) = \lambda_{\max}(B^*A^*AB) = \lambda_{\max}(BB^*A^*A) \leq \lambda_{\max}(BB^*)\lambda_{\max}(A^*A) = \sigma^2_{\max}(A)\sigma^2_{\max}(B)$.

To show the inequality for sum, let $\tilde{A}=\begin{pmatrix} 0 & A \\ A^* & 0 \end{pmatrix}$ and $\tilde{B}=\begin{pmatrix} 0 & B \\ B^* & 0 \end{pmatrix}$. Then $\tilde{A}$ and $\tilde{B}$ are Hermitian matrices. The largest eigenvalue of $\tilde{A}+\tilde{B}$ is $\sigma_{\max}(A+B)$, the largest eigenvalue of $\tilde{A}$ is $\sigma_{\max}(A)$, and that of $\tilde{B}$ is $\sigma_{\max}(B)$. Since $\lambda_{\max}(\tilde{A}+\tilde{B}) \leq \lambda_{\max}(\tilde{A}) + \lambda_{\max}(\tilde{B})$, the desired inequality follows.

For the last inequality, it is sufficient to notice that

$$A^2 - B^2 = \frac{1}{2}(A+B)(A-B) + \frac{1}{2}(A-B)(A+B).$$

4.55 By computing A^*A directly, we see that the singular values of A are $\sqrt{n}, \sqrt{n}, 1, \ldots, 1$. Since A is real symmetric, the singular values are the absolute values of the eigenvalues of A. Notice that $\operatorname{tr} A = -(n-2)$. Since A is congruent to $\begin{pmatrix} 1 & 0 \\ 0 & -I_{n-1}-e^te \end{pmatrix}$, where $e = (1, \ldots, 1)$ is a row vector of $n-1$ components, A has $n-1$ negative eigenvalues. The eigenvalues of A are $\sqrt{n}, -\sqrt{n}, -1, \ldots, -1$.

4.56 The fact that $|A+B| \geq |A|$ for $A > 0$ and $B \geq 0$ will play the basic role in the proofs. Note that equality holds if and only if $B = 0$.

(a) By induction on n. Let $A = \begin{pmatrix} a_{11} & \alpha \\ \alpha^* & A_1 \end{pmatrix}$. Then $A_1 \geq 0$. If $|A_1| = 0$, then $|A| = 0$. If $a_{11} = 0$, then $\alpha = 0$. There is nothing to show in both cases. Assume that $a_{11} \neq 0$. Upon computation,

$$\begin{aligned} |A| &= \begin{vmatrix} a_{11} & 0 \\ 0 & A_1 - a_{11}^{-1}\alpha^*\alpha \end{vmatrix} \\ &= a_{11}|A_1 - a_{11}^{-1}\alpha^*\alpha| \\ &\leq a_{11}|A_1|, \end{aligned}$$

since $0 \leq A_1 - a_{11}^{-1}\alpha^*\alpha \leq A_1$. The desired inequality follows by induction on A_1. Equality case is readily seen.

(b) If B and D are both singular, there is nothing to prove. Without loss of generality, assume that B is nonsingular. Then

$$\begin{aligned} |A| &= \begin{vmatrix} B & C \\ C^* & D \end{vmatrix} \\ &= \begin{vmatrix} B & 0 \\ 0 & D - C^*B^{-1}C \end{vmatrix} \\ &= |B||D - C^*B^{-1}C| \\ &\leq |B||D|. \end{aligned}$$

If equality holds and if $D > 0$, then

$$C^*B^{-1}C = 0, \quad \text{or} \quad C = 0.$$

(c) If B^{-1} exists, then

$$C^*B^{-1}C \leq D \;\Rightarrow\; |C^*||B^{-1}||C| \leq |D| \;\Rightarrow\; |C^*C| \leq |B||D|.$$

If B is singular, using $B + \epsilon I, \epsilon > 0$ to substitute B, we have $|C^*C| \leq |B + \epsilon I||D|$. Letting $\epsilon \to 0$, the inequality follows. For the equality case, let $D - C^*B^{-1}C = E \geq 0$. If $E \neq 0$, then

$$|D| = |E + C^*B^{-1}C| > |C^*B^{-1}C|.$$

When B, C, and D are of different sizes, it is not true. Take

$$B = \begin{pmatrix} 1 & 0 \\ 0 & 0 \end{pmatrix}, \quad C = (1,0)^t, \quad D = 1.$$

(d) Use (a) on E^*E.

(e) Use (b) on F^*F.

(f) Apply (c) to

$$\begin{pmatrix} I & X^* \\ Y & I \end{pmatrix}^* \begin{pmatrix} I & X^* \\ Y & I \end{pmatrix} = \begin{pmatrix} I + Y^*Y & X^* + Y^* \\ X + Y & I + XX^* \end{pmatrix}.$$

4.57 If $A^2 \geq 0$, then the eigenvalues of A^2 are nonnegative. However, the eigenvalues of A^2 are the squares of the eigenvalues of A: If $\lambda^2 \geq 0$, then λ is real. The converse is not true in general.

4.58 (a) $H = H^* = A^t - iB^t$ implies $A = A^t$ and $B = -B^t$. Thus A is real symmetric and B is skew-symmetric. To show that $A \geq 0$, we show that the eigenvalues of A are all nonnegative.

Let U be an orthogonal matrix such that U^tAU is the real diagonal matrix with the eigenvalues of A on the diagonal. Since

$$(U^tBU)^t = U^tB^tU = -U^tBU,$$

the diagonal entries of U^tBU are all zero, and so are the diagonal entries of iU^tBU. However,

$$U^tHU = U^tAU + iU^tBU \geq 0,$$

thus the diagonal entries of H, which are the eigenvalues of A, are all nonnegative.

(b) Consider the 2×2 minor of H formed by the entries on the s-th and t-th columns and rows.

(c) Following the proof of (a) and using the Hadamard inequality

$$|H| = |U^tHU| = |U^tAU + iU^tBU| \leq |U^tAU| = |A|.$$

Equality holds if and only if $B = 0$ or A has a zero eigenvalue.

(d) It is immediate from (c).

The converse of (d) is not true. Consider

$$H = \begin{pmatrix} 1 & 0 \\ 0 & 1 \end{pmatrix} + i \begin{pmatrix} 0 & 1 \\ -1 & 0 \end{pmatrix}.$$

If H is just Hermitian, then $B^t = -B$. Other parts are inconclusive.

4.59 (a) Let $A = UDV$ be a singular value decomposition of A. Then $A^*A = V^*DU^*UDV = V^*D^2V$, while $AA^* = UDVV^*DU^* = UD^2U^* = UV(V^*D^2V)V^*U^* = W(A^*A)W^*$, where $W = UV$ is a unitary matrix. So A^*A and AA^* are unitarily similar.

(b) $A = UDV = (UDU^*)(UV) = HP$, where $H = UDU^*$, $P = UV$.

(c) Let $A = UDV$, where U is $m \times m$, D is $m \times n$ with (i,i)-entries $s_i(A)$, the singular values of A, $i = 1, 2, \ldots, \min\{m, n\}$, V is $n \times n$. Then $AA^* = UDD^tU^*$ and $A = UDV = AA^*Q$, where $Q = URV$, R is the $m \times n$ matrix with (i,i)-entries $1/s_i(A)$ when $s_i(A) > 0$, and 0 otherwise.

4.60 (a) $[\det \mathrm{m}(A)]^2 = \det(A^*A) = |\det A|^2$.

(b) Since $A^* = A$.

(c) $A^*A = V^*D^2V$, so $\mathrm{m}(A) = V^*DV$. Similarly $\mathrm{m}(A^*) = UDU^*$.

(d) It is immediate from (c).

(e) The square root is unique.

(f) Upon computation

$$\begin{pmatrix} \mathrm{m}(A) & A^* \\ A & \mathrm{m}(A^*) \end{pmatrix} = \begin{pmatrix} V^* & 0 \\ 0 & U \end{pmatrix} \begin{pmatrix} D & D \\ D & D \end{pmatrix} \begin{pmatrix} V & 0 \\ 0 & U^* \end{pmatrix} \geq 0.$$

(g) Take $A = \left(\begin{smallmatrix} 0 & 1 \\ 2 & 0 \end{smallmatrix}\right)$.

(h) Since H is Hermitian, we show that H commutes with A^*A. $A^*AH = A^*HA = (HA)^*A = (AH)^*A = HA^*A$. It follows that H commutes with the square root of A^*A; that is, $\mathrm{m}(A)$.

By direct computation, $\mathrm{m}(A)$ and $\mathrm{m}(A^*)$ are, respectively,

$$\mathrm{m}(A) = \begin{pmatrix} 0 & 0 \\ 0 & 1 \end{pmatrix}, \quad \mathrm{m}(A^*) = \begin{pmatrix} 1 & 0 \\ 0 & 0 \end{pmatrix};$$

$$\mathrm{m}(A) = \mathrm{m}(A^*) = I_2;$$

$$\mathrm{m}(A) = \frac{1}{\sqrt{2}} \begin{pmatrix} 1 & 1 \\ 1 & 1 \end{pmatrix}, \quad \mathrm{m}(A^*) = \begin{pmatrix} \sqrt{2} & 0 \\ 0 & 0 \end{pmatrix}.$$

4.61 Note that $(A,B)^*(A,B) \geq 0$. For the second part, use Problem 4.56. A direct proof goes as follows. If $r(A) < n$, then

$$|A^*A| = |A^*B| = |B^*A| = 0.$$

Otherwise, observing that $\lambda_{\max}[A(A^*A)^{-1}A^*] = 1$, we have

$$I_m \geq A(A^*A)^{-1}A^*,$$

which implies

$$B^*B \geq B^*A(A^*A)^{-1}A^*B.$$

Taking the determinants,

$$\begin{aligned} |B^*B| &\geq |B^*A(A^*A)^{-1}A^*B| \\ &= |B^*A||(A^*A)^{-1}||A^*B| \\ &= |B^*A||A^*A|^{-1}|A^*B|. \end{aligned}$$

$\begin{pmatrix} A^*A & B^*A \\ A^*B & B^*B \end{pmatrix} \not\geq 0$ in general. But $\begin{pmatrix} |A^*A| & |B^*A| \\ |A^*B| & |B^*B| \end{pmatrix} = \begin{pmatrix} |A^*A| & |A^*B| \\ |B^*A| & |B^*B| \end{pmatrix}^t \geq 0$.

4.62 If $A = U(I_r \oplus 0)U^*$ for some unitary matrix U, then it is easy to check that $\frac{A+A^*}{2} = AA^*$. Now suppose $\frac{A+A^*}{2} = AA^*$ and let $A = U^*DU$, where U is unitary and D is an upper-triangular matrix with main diagonal entries $\lambda_1, \lambda_2, \ldots, \lambda_n$. Then $\frac{A+A^*}{2} = AA^*$ is the same as $\frac{D+D^*}{2} = DD^*$. The (1,1)-entry of $\frac{D+D^*}{2}$ is $\frac{\lambda_1+\bar{\lambda}_2}{2}$. Computing the (1,1)-entry of DD^*, we see that λ_1 must be a nonnegative number and the first row of D contains only 0 other than the (1,1)-entry λ_1. Then $\lambda_1 = \lambda_1^2$ gives $\lambda_1 = 0$ or 1. Inductively, we see that D is a diagonal matrix with entries on the main diagonal are either 0 or 1.

4.63 It suffices to note that BB^* is invertible for $r(BB^*) = r(B)$ and that

$$\begin{pmatrix} A \\ B \end{pmatrix} (A^*, B^*) = \begin{pmatrix} AA^* & AB^* \\ BA^* & BB^* \end{pmatrix} \geq 0.$$

4.64 If $|\lambda A - B| = 0$, then $0 = |A^{-\frac{1}{2}}|\,|\lambda A - B|\,|A^{-\frac{1}{2}}| = |\lambda I - A^{-\frac{1}{2}}BA^{-\frac{1}{2}}|$. Since $B > 0$, $A^{-\frac{1}{2}}BA^{-\frac{1}{2}} > 0$, thus the eigenvalue of $A^{-\frac{1}{2}}BA^{-\frac{1}{2}}$ are all positive. Hence $\lambda > 0$. If all the roots of $|\lambda A - B| = 0$ are 1, then all the roots of $|\lambda I - A^{-\frac{1}{2}}BA^{-\frac{1}{2}}|$ are 1. Thus $A^{-\frac{1}{2}}BA^{-\frac{1}{2}} = I$. Therefore, $A = B$. Conversely, if $A = B$, then $|\lambda A - B| = |\lambda A - A| = (\lambda - 1)^n|A|$. Since $|A| \neq 0$, $\lambda = 1$.

4.65 (a) $\operatorname{diag}(a_{11}, \ldots, a_{nn})$.

(b) A.

(c) One way to prove this is to write $A = U^*PU = \sum_i \lambda_i u_i^* u_i$ and $B = V^*QV = \sum_j \mu_j v_j^* v_j$, where λ_i and μ_j are nonnegative numbers, and u_i and v_j are rows of the unitary matrices P and Q, respectively. Then directly compute $x^*(A \circ B)x$, where $x \in \mathbb{C}^n$. Another way to prove it is to use tensor product. Since $A, B \geq 0$, suppose that $A = C^*C = ((c_i, c_j))$ and $B = D^*D = ((d_i, d_j))$, where c_i, d_i be the i-th columns of C and D, respectively. Let

$$c_i \otimes d_j = (c_{1i}d_{1j}, \ldots, c_{1i}d_{nj}, \ldots, c_{ni}d_{1j}, \ldots, c_{ni}d_{nj})^t.$$

It is easy to check that

$$A \circ B = ((c_i, c_j)(d_i, d_j)) = ((c_i \otimes d_i, c_j \otimes d_j)) = K^*K,$$

where K is the matrix with i-th column $c_i \otimes d_i$.

(d) Note that $\lambda_{\max}(A)I - A \geq 0$. Thus

$$[\lambda_{\max}(A)I - A] \circ B \geq 0 \quad \text{or} \quad \lambda_{\max}(A)(I \circ B) \geq A \circ B.$$

The conclusion then follows since $\max_i\{b_{ii}\} \leq \lambda_{\max}(B)$.

(e) No. Take $A = \left(\begin{smallmatrix} 1 & 1 \\ 1 & 1 \end{smallmatrix}\right)$, $B = \left(\begin{smallmatrix} 1 & 0 \\ 0 & 1 \end{smallmatrix}\right)$.

(f) Note that $a_{ii}b_{ii} \leq \frac{1}{2}(a_{ii}^2 + b_{ii}^2)$.

4.66 Compute the corresponding entries on both sides.

4.67 Let $M = \begin{pmatrix} A & A^{1/2} \\ A^{1/2} & I \end{pmatrix}$, $N = \begin{pmatrix} I & B^{1/2} \\ B^{1/2} & B \end{pmatrix}$. Then M, $N \geq 0$ and

$$\begin{aligned} M \circ N &= \begin{pmatrix} A \circ I & A^{1/2} \circ B^{1/2} \\ A^{1/2} \circ B^{1/2} & B \circ I \end{pmatrix} \\ &= \begin{pmatrix} I & A^{1/2} \circ B^{1/2} \\ A^{1/2} \circ B^{1/2} & I \end{pmatrix} \geq 0. \end{aligned}$$

4.68 (a) If $A > 0$, then $|A| > 0$ and $A^{-1} > 0$. For $x \neq 0$,

$$\begin{vmatrix} A & x \\ x^* & 0 \end{vmatrix} = \begin{vmatrix} A & 0 \\ 0 & -x^*A^{-1}x \end{vmatrix} = -|A|(x^*A^{-1}x) < 0.$$

If A is singular, then use $A + \epsilon I$, $\epsilon > 0$, for A above. Then

$$\begin{vmatrix} A + \epsilon I & x \\ x^* & 0 \end{vmatrix} = -|A + \epsilon I|(x^*A^{-1}x) < 0.$$

Letting $\epsilon \to 0$ yields that the determinant is 0 or negative.

(b) Denote $\delta = x^* A^{-1} x$. The inverse is

$$\begin{pmatrix} A^{-1}(I - \delta^{-1} x x^* A^{-1}) & \delta^{-1} A^{-1} x \\ \delta^{-1} x^* A^{-1} & -\delta^{-1} \end{pmatrix}.$$

4.69 (a) Notice that

$$\begin{pmatrix} I & 0 \\ -C^* B^{-1} & I \end{pmatrix} \begin{pmatrix} B & C \\ C^* & D \end{pmatrix} \begin{pmatrix} I & -B^{-1}C \\ 0 & I \end{pmatrix}$$

$$= \begin{pmatrix} B & 0 \\ 0 & D - C^* B^{-1} C \end{pmatrix}.$$

Thus

$$\begin{pmatrix} B & C \\ C^* & D \end{pmatrix}^{-1} = \begin{pmatrix} I & -B^{-1}C \\ 0 & I \end{pmatrix} \times$$

$$\begin{pmatrix} B^{-1} & 0 \\ 0 & (D - C^* B^{-1} C)^{-1} \end{pmatrix} \begin{pmatrix} I & 0 \\ -C^* B^{-1} & I \end{pmatrix}.$$

It follows by a direct computation that

$$U = B^{-1} + B^{-1} C (D - C^* B^{-1} C)^{-1} C^* B^{-1}$$

and

$$W = (D - C^* B^{-1} C)^{-1}.$$

Similarly, with D in the role of B,

$$W = D^{-1} + D^{-1} C^* (B - C D^{-1} C^*)^{-1} C D^{-1}$$

and

$$U = (B - C D^{-1} C^*)^{-1}.$$

(b) By (a)

$$\begin{pmatrix} B - U^{-1} & C \\ C^* & D \end{pmatrix} = \begin{pmatrix} C D^{-1} C^* & C \\ C^* & D \end{pmatrix} =$$

$$\begin{pmatrix} C & 0 \\ 0 & I \end{pmatrix} \begin{pmatrix} D^{-1} & I \\ I & D \end{pmatrix} \begin{pmatrix} C^* & 0 \\ 0 & I \end{pmatrix} \geq 0.$$

4.70 2 and 0, n copies of each.

4.71 (a) Note that

$$A^2 - 2A + I = (A-I)^2 = (A-I)^*(A-I) \geq 0.$$

Thus $A^2 + I \geq 2A$. Pre- and postmultiplying by $A^{-\frac{1}{2}}$, one has

$$A + A^{-1} \geq 2I.$$

It can also be proved by writing A as $A = U^* \operatorname{diag}(\lambda_1, \lambda_2, \ldots, \lambda_n)U$.

(b) Partition A and A^{-1} as

$$A = \begin{pmatrix} a & \alpha \\ \alpha^* & A_1 \end{pmatrix}, \quad A^{-1} = \begin{pmatrix} b & \beta \\ \beta^* & B_1 \end{pmatrix}.$$

Then by Problem 4.69

$$\left[A - \begin{pmatrix} \frac{1}{b} & 0 \\ 0 & 0 \end{pmatrix}\right] \circ \left[A^{-1} - \begin{pmatrix} 0 & 0 \\ 0 & A_1^{-1} \end{pmatrix}\right] \geq 0,$$

which yields

$$A \circ A^{-1} \geq \begin{pmatrix} 1 & 0 \\ 0 & A_1 \circ A_1^{-1} \end{pmatrix}.$$

The desired result follows by an induction on A_1.

4.72 (a) Let $x = (x_1, \ldots, x_n)^t$ be an eigenvector of A belonging to eigenvalue λ, and let

$$|x_i| = \max\{|x_1|, \ldots, |x_n|\}.$$

Consider the i-th component of both sides of $Ax = \lambda x$.

(b) $Ae = e$, where $e = (1, \ldots, 1)^t$.

(c) $Ae = e$ results in $A^{-1}e = e$ if A^{-1} exists.

4.73 Since A is a real orthogonal matrix; that is, $A^tA = I$, we see $|\lambda| = 1$. So $a^2 + b^2 = 1$. $A(x+yi) = (a+bi)(x+yi)$ implies $Ax = ax - by$ and $Ay = ay + bx$. Thus, $x^tA^t = ax^t - by^t$ and $y^tA^t = ay^t + bx^t$. Since $A^tA = I$, we have $x^tx = (ax^t - by^t)(ax - by) = a^2x^tx + b^2y^ty - 2abx^ty$. Because $a^2 + b^2 = 1$, we obtain $b^2x^tx = b^2y^ty - 2abx^ty$, which implies $2ax^ty = -bx^tx + by^ty$, as $b \neq 0$. With this in mind, compute x^ty:

$$\begin{aligned}
x^ty &= (ax^t - by^t)(ay + bx) \\
&= a^2x^ty + abx^tx - aby^ty - b^2y^tx \\
&= a^2x^ty + abx^tx - aby^ty - b^2x^ty \\
&= (a^2 - b^2)x^ty - a(-bx^tx + by^ty) \\
&= (a^2 - b^2)x^ty - 2a^2x^ty \\
&= -x^ty.
\end{aligned}$$

Thus $x^ty = 0$. Therefore $0 = -bx^tx + by^ty$ and if $b \neq 0$, $x^tx = y^ty$.

4.74 (a), (b), and (c) are easy.

(d) Let $Ux = \lambda x$, $x \neq 0$. Then $|\lambda| = 1$ because

$$|\lambda|^2 x^*x = (\lambda x)^*(\lambda x) = (Ux)^*(Ux) = x^*U^*Ux = x^*x.$$

(e) $Ux = \lambda x$ implies that $U^*x = \frac{1}{\lambda}x$.

(f) Let $U = V^* \operatorname{diag}(\lambda_1, \ldots, \lambda_n)V$, where V is a unitary matrix and the λ's are the eigenvalues of U, each of which equals 1 in absolute value. Let $y = Vx = (y_1, \ldots, y_n)^t$. Then y is a unit vector and

$$\begin{aligned} |x^*Ux| &= |\lambda_1|y_1|^2 + \cdots + \lambda_n|y_n|^2| \\ &\leq |\lambda_1||y_1|^2 + \cdots + |\lambda_n||y_n|^2 \\ &= |y_1|^2 + \cdots + |y_n|^2 = 1. \end{aligned}$$

(g) $\|Ux\| = \sqrt{x^*U^*Ux} = \sqrt{x^*x} = 1$.

(h) Each column or row vector of U is a unit vector.

(i) Let $Ux = \lambda_1 x$, $Uy = \lambda_2 y$, $\lambda_1 \neq \lambda_2$. Then

$$(\overline{\lambda_1}\lambda_2)(x^*y) = (\lambda_1 x)^*(\lambda_2 y) = (Ux)^*(Uy) = x^*y,$$

which, with $\lambda_1 \neq \lambda_2$ and $|\lambda_1| = |\lambda_2| = 1$, implies $x^*y = 0$.

(j) The column vectors form a basis since U is nonsingular. They form an orthonormal basis since $u_j^*u_i = 1$ if $i = j$ and 0 otherwise.

(k) Note that the k rows are linearly independent. Thus the rank of the submatrix of these rows is k. So there is a $k \times k$ submatrix whose determinant is nonzero.

(l) It may be assumed that A is diagonal. Note that each $|u_{ii}| \leq 1$.

(a), (b), (c), (g), and (j) imply that U is unitary.

4.75 By definition and direct verification.

4.76 Let $u_1, u_2, \ldots, u_n$ be the columns of U. Consider the matrix U^*U whose (i,j)-entry is $u_i^*u_j$ and use the Hadamard inequality.

4.77 Use induction on n. Suppose that A is upper-triangular. It can be seen by taking $x = (0, \ldots, 0, 1)^t$ that everything except the last component in the last column of A is 0.

4.78 Verify that $U^*U = I$. Note that for any positive integer k,

$$1 + \omega^k + \omega^{2k} + \cdots + \omega^{(n-1)k} = 0.$$

4.79 It is easy to compute that

$$\operatorname{tr}[(U \circ A)^*(U \circ A)] = \sum_{i,\, j=1}^{n} |u_{ij}|^2 |a_{ij}|^2.$$

Thus

$$\sigma_{\max}(U \circ A) \le \Big(\sum_{i,\, j=1}^{n} |u_{ij}|^2 |a_{ij}|^2 \Big)^{\frac{1}{2}}.$$

Now take U to be the unitary matrix in Problem 4.78.

4.80 Note that $\pm i$ cannot be the eigenvalues of A.

4.81 If A and A^* commute, then their inverses commute. Thus

$$(A^{-1}A^*)^*(A^{-1}A^*) = A(A^*)^{-1}A^{-1}A^* = AA^{-1}(A^*)^{-1}A^* = I.$$

So $A^{-1}A^*$ is unitary. Conversely, if $A^{-1}A^*$ is unitary, $(A^{-1}A^*)^* = (A^{-1}A^*)^{-1}$. Thus $A(A^*)^{-1} = (A^*)^{-1}A$, which yields $AA^* = A^*A$.

4.82 $\begin{pmatrix} \cos\theta & \sin\theta \\ -\sin\theta & \cos\theta \end{pmatrix}$, $\begin{pmatrix} \cos\theta & \sin\theta \\ \sin\theta & -\cos\theta \end{pmatrix}$, $\theta \in \mathbb{R}$.

4.83 Since A^t equals the adjoint $\operatorname{adj}(A)$ of A, we have $AA^t = A\operatorname{adj}(A) = |A|I_3$. It follows that $|A|^2 = |A|^3$. So $|A| = 0$ or $|A| = 1$. However, $a_{11} \neq 0$ and $|A| = a_{11}C_{11} + a_{12}C_{12} + a_{13}C_{13} = a_{11}^2 + a_{12}^2 + a_{13}^2 > 0$, where C_{ij} are cofactors of a_{ij}. So $|A| = 1$ and A is orthogonal.

4.84 First, find the eigenvalues of A and the corresponding eigenvectors, then find an orthonormal basis for $\mathbb{R}^4$ from the eigenvectors.

$$T = \begin{pmatrix} \frac{1}{\sqrt{2}} & -\frac{1}{\sqrt{6}} & \frac{1}{\sqrt{3}} \\ -\frac{1}{\sqrt{2}} & -\frac{1}{\sqrt{6}} & \frac{1}{\sqrt{3}} \\ 0 & \frac{2}{\sqrt{6}} & \frac{1}{\sqrt{3}} \end{pmatrix}, \quad T^tAT = \operatorname{diag}(2, 2, 8).$$

4.85 Suppose for real orthogonal matrices A and B, $A^2 = AB + B^2$. Then $A^2 = (A+B)B$ and $A(A-B) = B^2$. So $A + B = A^2B^{-1}$ and $A - B = A^{-1}B^2$ are orthogonal, since A^{-1} and B^2 are orthogonal. This reveals

$$I = (A+B)^t(A+B) = 2I + A^tB + B^tA$$

and

$$I = (A - B)^t(A - B) = 2I - A^tB - B^tA.$$

Adding them gives $2I = 4I$. A contradiction. If A and B are assumed to be invertible, then it is possible. Take $A = I$ and $B = \lambda I$, where λ is a root of $\lambda^2 + \lambda - 1 = 0$, say, $\lambda = \frac{-1+\sqrt{5}}{2}$, for instance.

4.86 First note that

$$|A + B| = |A^t + B^t|.$$

Multiply both sides by $|A|$ and $|B|$, respectively, to get

$$|A||A + B| = |A||A^t + B^t| = |I + AB^t|$$

and

$$|B||A + B| = |BA^t + I|^t = |I + AB^t|.$$

Thus

$$(|A| - |B|)|A + B| = 0$$

and

$$|A| = |B| \quad \text{if} \quad |A + B| \neq 0,$$

which implies $|A| = 0$, as $|A| + |B| = 0$. Thus A is singular. This contradicts the orthogonality of A.

It is false for unitary matrices. Take $A = I_2$ and $B = iI_2$.

4.87 Since $A > 0$, there exists a real invertible matrix P such that $P^tAP = I$. Since $(P^tBP)^t = -P^tBP$, P^tBP is real skew-symmetric and thus its eigenvalues are 0 or nonreal complex numbers; the nonreal eigenvalues appear in conjugate pairs. Let T be a real invertible matrix such that $T^{-1}(P^tBP)T = \operatorname{diag}(\lambda_1, \lambda_2, \ldots, \lambda_n)$, where the λ_i are either 0 or nonreal complex numbers in conjugate pairs. It follows that $T^{-1}P^t(A + B)PT = \operatorname{diag}(1 + \lambda_1, 1 + \lambda_2, \ldots, 1 + \lambda_n)$. By taking determinants, we see that $|A + B| > 0$.

4.88 By definition and direct verification.

4.89 $A = B + C = F + iG$, where $B = F = \frac{A+A^*}{2}$ is Hermitian, $C = \frac{A-A^*}{2}$ is skew-Hermitian, and $G = -i\frac{A-A^*}{2}$ is Hermitian.

4.90 (a) No.

(b) Yes.

(c) No.

4.91 It is easy to see that (a)$\Leftrightarrow$(b). We first show that (a), (c), (d), and (e) are all equivalent. To see (a) implies (c), use induction. If $n = 1$, there is nothing to prove. Suppose it is true for $n - 1$. For the case of n, let u_1 be a unit eigenvector belonging to the eigenvalue λ_1 of A. Let U_1 be a unitary matrix with u_1 as its first column. Then $U_1^* A U_1$ is of the form

$$\begin{pmatrix} \lambda_1 & \alpha \\ 0 & A_1 \end{pmatrix}.$$

The normality of A yields $\alpha = 0$. (c) follows by induction.

It is obvious that (c) implies (a).

(c)$\Leftrightarrow$(d): Note that $U^*AU = D$, where $D = \operatorname{diag}(\lambda_1, \ldots, \lambda_n)$, if and only if $AU = UD$, or $AU_i = \lambda_i U_i$, where U_i is the i-th column of U.

(c)$\Leftrightarrow$(e): If $Av = \lambda v$, $v \neq 0$, assume that v is a unit vector. Let $U = (v, U_1)$ be a unitary matrix. Then, since A is normal,

$$U^*AU = \begin{pmatrix} \lambda & 0 \\ 0 & A_1 \end{pmatrix}.$$

It is easy to see by taking conjugate transpose that v is an eigenvector of A^* corresponding to $\overline{\lambda}$. To see the other direction, let A be an upper-triangular matrix with nonzero (1,1)-entry. Take $e_1 = (1, 0, \ldots, 0)^t$. Then e_1 is an eigenvector of A. If e_1 is an eigenvector of A^*, then the first column of A^* must consist of zeros except the first component. Use induction hypothesis on n.

(f)$\Leftrightarrow$(c): If $A^* = AU$, then

$$A^*A = A^*(A^*)^* = (AU)(AU)^* = AA^*$$

and A is normal; hence (c) holds. To see the converse, let

$$A = S^* \operatorname{diag}(\lambda_1, \ldots, \lambda_n) S,$$

where S is unitary. Take

$$U = S^* \operatorname{diag}(l_1, \ldots, l_n) S,$$

where $l_i = \frac{\overline{\lambda_i}}{\lambda_i}$ if $\lambda_i \neq 0$, and $l_i = 1$ otherwise, $i = 1, \ldots, n$.

Similarly (g) is equivalent to (c).

(c)$\Rightarrow$(h) is obvious. To see the converse, assume that A is an upper-triangular matrix and consider the trace of A^*A.

(i)⇒(c): Let A be upper-triangular. Consider the diagonal of A^*A.

(j)⇒(a): Note that for matrices X and Y of the same size

$$\operatorname{tr}(XY) = \operatorname{tr}(YX).$$

On one hand, by computation,

$$\operatorname{tr}(A^*A - AA^*)^*(A^*A - AA^*) = \operatorname{tr}(A^*A - AA^*)^2 =$$

$$\operatorname{tr}(A^*A)^2 - \operatorname{tr}[(A^*)^2A^2] - \operatorname{tr}[A^2(A^*)^2] + \operatorname{tr}(AA^*)^2 = 0.$$

On the other hand,

$$\operatorname{tr}(X^*X) = 0 \Leftrightarrow X = 0,$$

thus $A^*A - AA^* = 0$; that is, A is normal.

(k)⇒(a): This is because $\|Ax\| = \|A^*x\|$ implies

$$x^*A^*Ax = x^*AA^*x;$$

that is

$$x^*(A^*A - AA^*)x = 0$$

for all $x \in \mathbb{C}^n$. Thus $A^*A - AA^* = 0$ and A is normal.

(l)⇒(a): By a direct verification.

(m)⇒(a): Note that $\operatorname{tr}(A^*A - AA^*) = 0$.

(n)⇒(j): If $AA^*A = A^*A^2$, then by multiplying A^* from the left

$$A^*AA^*A = (A^*)^2A^2.$$

Thus (j) is immediate by taking trace.

(o)⇒(a): We show that (o)⇒(j). Since A commutes with $AA^* - A^*A$,

$$A^2A^* + A^*A^2 = 2AA^*A.$$

Multiply both sides by A^* from the left

$$A^*A^2A^* + (A^*)^2A^2 = 2A^*AA^*A.$$

(j) follows by taking trace for both sides.

4.92 (a) Take $B = \frac{A+A^*}{2}$ and $C = -i\frac{A-A^*}{2}$. Then $BC = CB$.

(b) Let $A = U\operatorname{diag}(\lambda_k)U^*$ and each $\lambda_k = |\lambda_k|e^{i\theta_k}$. Take $H = D\operatorname{diag}(|\lambda_k|)U^*$ and $P = U\operatorname{diag}(e^{i\theta})U^*$. Then $A = HP = PH$.

The converses of (a) and (b) are also true.

4.93 B and D are normal, $C = 0$.

4.94 Denote the correspondence by $M \sim N$. It is easy to show that if $M_1 \sim N_1$ and $M_2 \sim N_2$ then $M_1M_2 \sim N_1N_2$ and $M^* \sim N^t$. When $M \geq 0$, write $M = C^*C$.

4.95 (a) We show that $\operatorname{Ker} A^* \subseteq \operatorname{Ker} A$. The other way is similar. Let

$$x \in \operatorname{Ker} A^* \quad \text{or} \quad A^*x = 0,$$

then $AA^*x = 0$ and $A^*Ax = 0$ since A is normal. Thus $x^*A^*Ax = (Ax)^*(Ax) = 0$ and $Ax = 0$; that is, $x \in \operatorname{Ker} A$.

(b) Let

$$x \in \operatorname{Im} A^* \quad \text{and} \quad x = A^*y.$$

Since A is normal, by Problem 4.91, assume $A^* = AU$ for some unitary matrix U, then

$$x = A^*y = AUy \in \operatorname{Im} A.$$

Thus $\operatorname{Im} A^* \subseteq \operatorname{Im} A$. The other way around is similar.

(c) Since $n = \dim(\operatorname{Im} A) + \dim(\operatorname{Ker} A)$ and $\operatorname{Im} A^* = \operatorname{Im} A$, we show

$$\operatorname{Im} A^* \cap \operatorname{Ker} A = \{0\}.$$

Let $x = A^*y$ and $Ax = 0$. Then

$$0 = y^*Ax = y^*AA^*y = (A^*y)^*(A^*y) \quad \Rightarrow \quad x = A^*y = 0.$$

4.96 First consider the case where A is a diagonal matrix. Let $A = \operatorname{diag}(\lambda_1, \ldots, \lambda_n)$. Then $AB = BA$ yields $\lambda_i b_{ij} = \lambda_j b_{ij}$; that is, $(\lambda_i - \lambda_j)b_{ij} = 0$. Thus $\overline{\lambda_i} b_{ij} = \overline{\lambda_j} b_{ij}$, which implies $A^*B = BA^*$. For the general case, let $A = U^*\operatorname{diag}(\lambda_1, \ldots, \lambda_n)U$ for some unitary matrix U, then use the above argument with UBU^* for B.

4.97 Necessity $\Rightarrow$: $A\bar{A} = 0 \Rightarrow A^*A\bar{A} = 0 \Rightarrow AA^*\bar{A} = 0 \Rightarrow \bar{A}A^tA = 0$. So $(A^tA)^*(A^tA) = A^*\bar{A}A^tA = 0$ and $A^tA = 0$.

Similarly, $A\bar{A} = 0 \Rightarrow \bar{A}A = 0 \Rightarrow \bar{A}AA^* = 0 \Rightarrow \bar{A}A^*A = 0 \Rightarrow AA^t\bar{A} = 0$. So $(AA^t)(AA^t)^* = AA^t\bar{A}A^* = 0$ and $AA^t = 0$.

Sufficiency $\Leftarrow$: $(A\bar{A})^*(A\bar{A}) = A^tA^*A\bar{A} = A^tAA^*\bar{A} = 0$. So $A\bar{A} = 0$.

4.98 (a) Let λ_1 be an eigenvalue of A and let

$$V_{\lambda_1} = \{ x \mid Ax = \lambda_1 x,\ x \neq 0 \}$$

be the eigenspace of λ_1. Since A and B commute, for $x \in V_{\lambda_1}$,

$$A(Bx) = B(Ax) = B(\lambda_1 x) = \lambda_1(Bx),$$

V_{λ_1} is an invariant subspace of $\mathbb{C}^n$ under B, regarded as a linear transformation on $\mathbb{C}^n$. As a linear transformation on V_{λ_1} over $\mathbb{C}$, B has an eigenvalue $\mu_1 \in \mathbb{C}$ and a unit eigenvector $u_1 \in V_{\lambda_1}$. Let U_1 be a unitary matrix whose first column is u_1, then

$$U_1^* A U_1 = \begin{pmatrix} \lambda_1 & \alpha \\ 0 & A_1 \end{pmatrix} \quad \text{and} \quad U_1^* B U_1 = \begin{pmatrix} \mu_1 & \beta \\ 0 & B_1 \end{pmatrix}.$$

The normality of A and B implies that $\alpha = \beta = 0$. Now apply induction hypotheses to A_1 and B_1.

(b) It is immediate from (a). If the condition $AB = BA$ is dropped, the conclusion does not necessarily follow. Take

$$A = \begin{pmatrix} 1 & 0 \\ 0 & 0 \end{pmatrix}, \quad B = \begin{pmatrix} 1 & 1 \\ 1 & 1 \end{pmatrix}.$$

(c) If $AB^* = B^*A$, then $BA^* = A^*B$. It follows that

$$\begin{aligned}
(AB)(AB)^* &= A(BB^*)A^* \\
&= (AB^*)(BA^*) \quad (B \text{ is normal}) \\
&= (B^*A)(A^*B) \\
&= B^*(AA^*)B \\
&= (B^*A^*)(AB) \quad (A \text{ is normal}) \\
&= (AB)^*(AB).
\end{aligned}$$

Hence AB is normal. Similarly, BA is normal.

(d) By a direct computation.

(e) Recall (Problem 4.91) that a matrix X is normal if and only if

$$\operatorname{tr}(X^*X) = \sum_{i=1}^{n} |\lambda_i(X)|^2.$$

We show that

$$\operatorname{tr}(A^*B^*BA) = \sum_{i=1}^{n} |\lambda_i(BA)|^2.$$

$$\begin{aligned}
\operatorname{tr}(A^*B^*BA) &= \operatorname{tr}(B^*BAA^*) \quad (\text{use } \operatorname{tr} XY = \operatorname{tr} YX) \\
&= \operatorname{tr}(BB^*A^*A) \quad (A, B \text{ are normal}) \\
&= \operatorname{tr}(B^*A^*AB) \quad (\text{use } \operatorname{tr} XY = \operatorname{tr} YX) \\
&= \sum_{i=1}^{n} |\lambda_i(AB)|^2 \quad (AB \text{ is normal}) \\
&= \sum_{i=1}^{n} |\lambda_i(BA)|^2.
\end{aligned}$$

$\left(\begin{smallmatrix} 1 & 1 \\ 0 & 0 \end{smallmatrix}\right)$ and $\left(\begin{smallmatrix} 1 & 0 \\ 1 & 0 \end{smallmatrix}\right)$ are not normal, but their product is normal.

4.99 A is diagonalizable with eigenvalues 1, 1, and -1 or 1, -1, and -1.

4.100 If λ is an eigenvalue of A, then $\lambda^3 + \lambda = 0$, which has only zeros 0, i, and $-i$. The complex roots of real-coefficient polynomial appear in conjugate pairs, so do the eigenvalues of a real matrix. Thus $\operatorname{tr} A = 0$.

4.101 The eigenvalues λ of A are k-th primitive roots of 1, so $\frac{1}{\lambda} = \bar{\lambda}$. Since $x^k = 1$ has at most k roots in $\mathbb{C}$, some λ_i's must be the same when $k < n$.

4.102 Use Jordan form of A. Let J be a Jordan block of A. Then $J^k = I$. J has to be 1×1. Thus A is diagonalizable. For B, the characteristic polynomial is $\lambda^2 + 1$, which has no real solution. So B is not diagonalizable over $\mathbb{R}$. (But B is diagonalizable over $\mathbb{C}$.)

4.103 Use Jordan form and consider the case $J^2 = J$ for Jordan block J.

4.104 Let $A^m = 0$ and $B^n = 0$. Then $AB = BA$ implies $(AB)^k = A^k B^k = 0$, where $k \geq \max\{m, n\}$. For $A + B$, expanding $(A + B)^{m+n}$, since $AB = BA$, we see that every term in the expansion contains $A^p B^q$, where $p + q = m + n$. So either $p \geq m$ or $q \geq n$. Thus $A^p B^q = 0$ and $(A + B)^{m+n} = 0$.

4.105 (a) Let $A^m = 0$. Then $I = I - A^m = (I - A)(I + A + A^2 + \cdots + A^{m-1})$. Thus $I - A$ is invertible and $(I - A)^{-1} = I + A + A^2 + \cdots + A^{m-1}$.

(b) Replace A by $-A$ in (a).

(c) Because all eigenvalues of A are 0.

(d) If A is diagonalizable, then $A = 0$ as all eigenvalues of A are 0.

4.106 Since $A^2 = A$ and $B^2 = B$, we have

$$(A+B)^2 = A^2 + AB + BA + B^2 = A + AB + BA + B.$$

If $(A+B)^2 = A+B$, then $AB + BA = 0$; that is, $AB = -BA$. Also,

$$AB = A^2B = A(AB) = A(-BA) = -(AB)A = BA^2 = BA.$$

It follows that $BA = -BA$ and $BA = 0$. So $AB = 0$.

If $AB = BA = 0$, then obviously $(A+B)^2 = A+B$.

4.107 When A and B are Hermitian, $(AB)^* = B^*A^* = BA$. Thus

$$(BA)^2 = [(AB)^*]^2 = [(AB)^2]^* = (AB)^* = B^*A^* = BA.$$

4.108 If $A^2 = A$, then there exists an invertible matrix T such that

$$A = T^{-1}\begin{pmatrix} I_r & 0 \\ 0 & 0 \end{pmatrix} T = T^{-1}\begin{pmatrix} I_r \\ 0 \end{pmatrix}(I_r, 0)T = BC,$$

where $B = T^{-1}\left(\begin{smallmatrix} I_r \\ 0 \end{smallmatrix}\right)$, $C = (I_r, 0)T$; both have rank r and $CB = I_r$. The converse is easy to check. If $A^2 = A$, let $T^{-1}AT = \operatorname{diag}(I_r, 0)$ for some T. So $T^{-1}(A+I)T = \operatorname{diag}(2I_r, I_{n-r})$ and $T^{-1}(2I - A)T = \operatorname{diag}(I_r, 2I_{n-r})$. Thus $|2I_n - A| = 2^{n-r}$ and $|A + I_n| = 2^r$.

4.109 Let T be an invertible matrix such that $T^{-1}AT = \operatorname{diag}(I_r, 0_{n-r})$, where $1 \le r \le n-1$. Take $B = T\operatorname{diag}(0, J)T^{-1}$, where J is the $k \times k$ Jordan block with main diagonal entries 0. Then one may verify that $AB = BA = 0$ and $(A+B)^{k+1} = (A+B)^k \neq (A+B)^{k-1}$.

4.110 Let r be the rank of A. Since $A^2 = A$, the eigenvalues of A are either 1 or 0. Using the Jordan form of A, by $A^2 = A$, we see that every Jordan block of A must be 1×1. Thus A is diagonalizable. It follows that (a) $|A + I| = 2^r$ and $|A - I| = 0$; (b) $r(A)$ equals the number of nonzero eigenvalues of A, which are 1's, and also equals $\operatorname{tr}(A)$; (c) $\dim(\operatorname{Im} A) = r(A) = \operatorname{tr}(A)$.

4.111 (a) Note that for every positive integer k,

$$AB = -BA \quad \Rightarrow \quad AB^{2k-1} = -B^{2k-1}A.$$

Thus if $B = AX + XA$, then

$$B^{2k} = (AX + XA)B^{2k-1} = A(XB^{2k-1}) - (XB^{2k-1})A,$$

which implies that $\operatorname{tr} B^{2k} = 0$ for all positive integers k; consequently, B^2 is nilpotent, and so is B, a contradiction.

(b) Without loss of generality, one may assume that A is the diagonal matrix $\operatorname{diag}(\lambda_1, \ldots, \lambda_n)$, all $\lambda_i > 0$. The existence and uniqueness of the equation can be checked directly by taking $x_{ij} = \frac{1}{\lambda_i+\lambda_j} b_{ij}$. To show that $X \geq 0$ when $B \geq 0$, we first note that X is Hermitian by taking the conjugate transpose for both sides and because the solution is unique. To show that the eigenvalues of X are all nonnegative, let λ be an eigenvalue of X and let u be a corresponding vector. Then $0 \leq u^*Bu = u^*XAu + u^*AXu = 2\lambda(u^*Au)$. Thus $\lambda \geq 0$, as $u^*Au > 0$.

4.112 $AX + XB = C$ has a unique solution if and only if $\begin{pmatrix} A & C \\ 0 & -B \end{pmatrix}$ and $\begin{pmatrix} A & 0 \\ 0 & -B \end{pmatrix}$ are similar via $\begin{pmatrix} I & X \\ 0 & -I \end{pmatrix}$.

4.113 Note that the idempotent Hermitian matrices are the positive semidefinite matrices with eigenvalues 1's and 0's, and that if A_1 is a principal submatrix of a positive semidefinite matrix A, then

$$\lambda_{\min}(A) \leq \lambda_{\min}(A_1) \leq \lambda_{\max}(A_1) \leq \lambda_{\max}(A).$$

Since $B \leq A \Leftrightarrow AB = B$ is equivalent to

$$U^*BU \leq U^*AU \quad \Leftrightarrow \quad (U^*AU)(U^*BU) = U^*BU,$$

where U is unitary, we may only consider the case in which $B = \begin{pmatrix} I_r & 0 \\ 0 & 0 \end{pmatrix}$, where $r = r(B)$ is the rank of B. Partition A conformably as

$$A = \begin{pmatrix} M_1 & M_2 \\ M_2^* & M_3 \end{pmatrix}.$$

To show $B \leq A \Rightarrow AB = B$, note that $A - B \geq 0$ implies $M_1 - I_r \geq 0$ and that the eigenvalues of M_1 are all equal to 1. It follows that $M_1 = I_r$. Thus $M_2 = 0$ and $A = \begin{pmatrix} I_r & 0 \\ 0 & M_3 \end{pmatrix}$, where $M_3 \geq 0$. It is immediate that $AB = B$.

For the other way around, $AB = B$ results in $M_2 = 0, M_1 = I_r$.

4.114 If $x \in \operatorname{Im} A$, let $x = Ay$. Then $Ax = A^2y = Ay = x$. Conversely, if $x = Ax$, then obviously $x \in \operatorname{Im} A$.

To show that $\operatorname{Im} A = \operatorname{Im} B$ implies $A = B$, we compute, for any v, $(A-B)^2v = A^2v + B^2v - ABv - BAv = Av + Bv - A(Bv) - B(Av) = 0$, since $A(Bv) = Bv$ and $B(Av) = Av$. $A - B$ is Hermitian, so $A = B$.

4.115 It is sufficient to show that (d)⇒(a). This is immediate since $A^2 = I$ and $A^*A = I$ imply $A^* = A$.

4.116 (a) By a direct verification.

(b) Write $A^2 - I = 0$ as $(A+I)(A-I) = 0$. Then

$$r(A+I) + r(A-I) \leq n.$$

However,

$$n = r(A) = r[(A+I) + (A-I)] \leq r(A+I) + r(A-I).$$

(c) If $Ax = \lambda x$, for some $x \neq 0$, then

$$x = Ix = A^2x = A(Ax) = \lambda Ax = \lambda^2 x \quad \Rightarrow \quad \lambda^2 = 1.$$

(d) For any $v \in V$,

$$v = \frac{1}{2}(v + Av) + \frac{1}{2}(v - Av) \in V_1 + V_{-1}.$$

Thus $V = V_1 + V_{-1}$. Note that $V_1 \cap V_{-1} = \{0\}$.

(e) Since $(A+I)(A-I) = 0$.

(a), (d), and (e) each imply that $A^2 = I$.

4.117 $ABA = B$ and $BAB = A$ imply $AB = BA^{-1}$ and $AB = B^{-1}A$. So $BA^{-1} = B^{-1}A$ and $A^2 = B^2$. $A = BAB = (ABA)(AB) = B^{-1}A^3B$. Thus $BA = A^3B$. But $BA = A^{-1}B$. We have $A^{-1}B = A^3B$, and $A^{-1} = A^3$; that is, $A^4 = I$. Similarly, $B^4 = I$.

4.118 Since $(A-B)(A+B) = (A+B)(B-A) = AB - BA$, we see that $\operatorname{Im}(AB - BA)$ is contained in both $\operatorname{Im}(A-B)$ and $\operatorname{Im}(A+B)$; that is, $\operatorname{Im}(AB - BA) \subseteq \operatorname{Im}(A-B) \cap \operatorname{Im}(A+B)$. For the converse, let $u \in \operatorname{Im}(A-B) \cap \operatorname{Im}(A+B)$ and write $u = (A+B)x = (A-B)y$. Then $(A-B)u = (AB-BA)x$ and $(A+B)u = (BA-AB)y$. By adding, $2Au = (AB-BA)(x-y)$. It follows that

$$u = A^2u = \frac{1}{2}A(AB - BA)(x-y) = \frac{1}{2}(AB - BA)A(y-x),$$

which is contained in $\operatorname{Im}(AB - BA)$.

4.119 If $A^2 = A$, then $\frac{1}{4}(B^2 + 2B + I) = \frac{1}{2}(B+I)$, which implies $B^2 = I$. If $B^2 = I$, then $A^2 = \frac{1}{4}(B^2 + 2B + I) = \frac{1}{4}(2I + 2B) = \frac{1}{2}(B+I) = A$.

4.120 Direct verification by definition.

4.121 (a) $n!$.

(b) By a direct verification. No.

(c) If P is a permutation matrix, then $PP^t = I$.

(d) Symmetric permutation matrices.

4.122 (a) By direct computations. Note that for any matrix A, AP is the matrix obtained from A by moving the last column of A to the first column.

(b) Consider

$$k_1 P + k_2 P^2 + \cdots + k_n P^n = 0.$$

Since the k's are in different positions of the matrix on the left-hand side, it follows that all the k's must be equal to zero.

(c) It is routine to check by definition. Note $P^t = P^{-1}$.

(d) $i + j$ is divisible by n.

(e) Note that the characteristic polynomial of P is $\lambda^n - 1$.

(f) Take $T = (t_{ij})$ to be the permutation matrix with $t_{k,k'} = 1$, $k = 1, 2, \ldots, n$, and 0 otherwise, where k' is the positive integer such that

$$0 < k' = (k-1)i + 1 - mn \le n$$

for some nonnegative integer m. For instance, if $n = 5$ and $i = 3$,

$$t_{11} = t_{24} = t_{32} = t_{45} = t_{53} = 1$$

and 0 otherwise. It is easy to check that $T^{-1} P^3 T = P$.

4.123 Let A be $k \times k$. Since A is invertible, there is at least one summand in the expansion of the determinant $\det A$ that is nonzero. For A is a matrix of integer entries, we may write A as $A = P + B$, where P is a permutation matrix and B is a nonnegative matrix. Notice that $P^k = I$. Thus in the expansion of $A^n = (P + B)^n$, there are infinitely many summands $P^i B P^j$ that are identical to B. If B had a nonzero (thus positive) entry, then the sum of the entries of A^n would be unbounded as $n \to \infty$. Therefore, $B = 0$ and $A = P$ is a permutation matrix. For the case of union of entries, because the powers of A collectively have only finitely many entries, we must have $A^m = A^n$ for some m and n. So $A^p = I$ for some p. Then expand $I = A^p = (P + B)^p = P^p + P^{p-1} B + \cdots$. The sum of the entries on the left-hand side is k, and then so is on the right. Thus $B = 0$.

4.124 Since every row and every column of A have one and only one nonzero entry, which is either 1 or -1, and all other entries are 0, it is easy to see that the powers of A: A^2, A^3, ..., have the same property; that is, for all positive integer m, every row and every column of A^m have one and only one nonzero entry, which is either 1 or -1, and all other entries are 0. Since there are a finite number of matrices that have the property, we have $A^p = A^q$ for some positive integers p and q, $p < q$. Since A is nonsingular, we have $A^{q-p} = I$.

4.125 A is an $(n-1) \times n$ matrix. Denote by B the submatrix of A by deleting the last column of A. Since the row sums of A are all zero, we may write $A = (B, BR)$, where $R = (-1, -1, \cdots, -1)^t$. Then

$$\begin{aligned} |AA^t| &= |(B, BR)(B, BR)^t| = |BB^t + BRR^tB^t| \\ &= |B(I + RR^t)B^t| = |B|^2\,|I + RR^t| = n|B|^2. \end{aligned}$$

4.126 Denote the rows of A by r_1, r_2, ..., r_n. Let S and T be the row and column indices of the $s \times t$ submatrix whose entries are all 1. Set $v = \sum_{i\in S} r_i$. Since the rows are mutually orthogonal, we have

$$\|v\|^2 = \sum_{i\in S,\, j\in T} r_j^t r_i = \sum_{i\in S} r_i^t r_i = sn.$$

However, the j-th coordinate of v is $\sum_{i\in S} a_{ij}$,

$$\|v\|^2 = \sum_{j=1}^{n} \left(\sum_{i\in S} a_{ij} \right)^2 \geq s^2 t.$$

It follows that $s^2 t \leq sn$ and this shows that $st \leq n$.

4.127 By direct computation, we have

$$(A + tB)^3 = (t^3 - 2)I_3, \quad B^3 = I_3, \quad A^3 = -2I_3.$$

4.128 Upon computation, $X^4 = (2pq)^2$, $Y^4 = (p^2 - q^2)^2$, $Z^4 = (p^2 + q^2)^2$.

Hints and Answers for Chapter 5

5.1 $\|\frac{1}{\|u\|}u\| = \frac{1}{\|u\|}\|u\| = 1$. So u is a unit vector. For $v, w \in V$,

$$\langle v, \langle v, w\rangle w\rangle = \overline{\langle v, w\rangle}\langle v, w\rangle = |\langle v, w\rangle|^2.$$

$\langle\langle v, w\rangle v, w\rangle = \langle v, w\rangle^2 \neq |\langle v, w\rangle|^2$ in general over $\mathbb{C}$.

5.2 The first identity is easy . For the second one, use $\langle u, \lambda v\rangle = \bar{\lambda}\langle u, v\rangle$.

5.3 (a) False. Take $u = (2, 2)$.

(b) True. By the Cauchy-Schwarz inequality.

(c) False. Take $u = v = (1, 1)$, $w = (1, -1)$.

(d) False. Take $u = v$, $w = -u$.

(e) True. By the triangle inequality.

5.4 (a) and (d) hold. (b) and (c) do not hold.

5.5 For $x = (x_1, x_2, x_3)^t$, $(x, x) = 2|x_2|^2 + |x_1|^2 + \bar{x_1}x_3 + x_1\bar{x_3} + 2|x_3|^2$. Since $|x_1|^2 + \bar{x_1}x_3 + x_1\bar{x_3} + |x_3|^2 \geq 0$, we see that $(x, x) \geq 0$. Equality holds if and only if $x = 0$. Other conditions are easy to verify. If the $(2, 2)$-entry is replaced with -2, then the positivity does not hold, since $((0, 1, 0), (0, 1, 0)) = -2$. So it's no longer an inner product. If x and y are switched on the right, then it's not an inner product.

5.6 It is routine to show by definition that V is a subspace of $\mathbb{R}^4$. To find a basis for V, note that x_3 and x_4 are free variables. Setting $x_3 = 1$, $x_4 = 0$ and $x_3 = 0$, $x_4 = 1$, respectively, we have a basis for V: $(1, 1, 1, 0)^t$ and $(1, -1, 0, 1)^t$. To find a basis for $V^\perp$, let $y = (y_1, y_2, y_3, y_4)^t \in V^\perp$. Then y is orthogonal to the basis vectors of V. So $y_1 + y_2 + y_3 = 0$ and $y_1 - y_2 + y_4 = 0$. This reveals $2y_1 = -y_3 - y_4$ and $2y_2 = y_4 - y_3$. It is easy to see that $(-1, 0, 1, 1)^t$ and $(-1, 1, 0, 2)^t$ form a basis for $V^\perp$ and $V^\perp = \{a(-1, 0, 1, 1)^t + b(-1, 1, 0, 2)^t \mid a, b \in \mathbb{R}\}$.

5.7 $\alpha_3 = (0, \frac{1}{\sqrt{2}}, -\frac{1}{\sqrt{2}}, 0)^t$, $\alpha_4 = (0, -\frac{1}{2}, -\frac{1}{2}, \frac{1}{\sqrt{2}})^t$.

5.8 $f_1(x) = 1 - \frac{x}{2}$, $f_2(x) = \frac{x}{2} - \frac{x^2}{3}$, $f_3(x) = \frac{x^2}{3} - \frac{x^3}{4}$ form a basis for $W^\perp$.

5.9 The first three are positive semidefinite. The last one is not. Take $u_1 = (0.1, 0.1, 0.1)^t, u_2 = (0.3, 0.7, 0.3)^t, u_3 = (0.4, 0.5, 0.7)^t$.

5.10 We need to show that $x^*Gx \geq 0$, where $x = (x_1, x_2, \ldots, x_n)^t$. Let $y = x_1u_1 + x_2u_2 + \cdots + x_nu_n$ and denote $g_{ij} = \langle u_j, u_i\rangle$. Then

$$0 \leq \langle y, y\rangle = \sum_{i,j=1}^{n} x_i\bar{x}_j\langle u_i, u_j\rangle = \sum_{i,j=1}^{n} x_i\bar{x}_j g_{ji} = x^*Gx.$$

G is singular if and only if $Gz = 0$ for some nonzero z. If $Gz = 0$ for some nonzero z, take $x = z$, we see that $y = 0$ and the vectors are linearly dependent. The converse is proved similarly. Note that $G \geq 0$ if and only if $G^t \geq 0$. So $H = G^t \geq 0$.

5.11 Since $\|u\| < 1, \|v\| < 1$, by the Cauchy-Schwarz inequality, we have $|\langle x, y\rangle| < 1$, where x, y are any choices of u and v. Recall the power series expansion $\frac{1}{1-r} = 1 + r + r^2 + r^3 + \cdots$ when $|r| < 1$.

$$\begin{aligned}\begin{pmatrix} \frac{1}{1-\langle u,u\rangle} & \frac{1}{1-\langle u,v\rangle} \\ \frac{1}{1-\langle v,u\rangle} & \frac{1}{1-\langle v,v\rangle} \end{pmatrix} &= \begin{pmatrix} \sum_k \langle u,u\rangle^k & \sum_k \langle u,v\rangle^k \\ \sum_k \langle v,u\rangle^k & \sum_k \langle v,v\rangle^k \end{pmatrix} \\ &= \sum_k \begin{pmatrix} \langle u,u\rangle^k & \langle u,v\rangle^k \\ \langle v,u\rangle^k & \langle v,v\rangle^k \end{pmatrix}.\end{aligned}$$

The positivity follows immediately because the sum of positive semidefinite matrices is again positive (with a bit more work on convergence) and because $\begin{pmatrix} a & b \\ \bar{b} & c \end{pmatrix} \geq 0 \Rightarrow \begin{pmatrix} a^k & b^k \\ \bar{b}^k & c^k \end{pmatrix} \geq 0$ for any positive integer k.

5.12 If there exist n vectors $u_1, u_2, \ldots, u_n \in \mathbb{C}^n$ such that $A = (a_{ij})$, where $a_{ij} = \langle u_j, u_i\rangle$, then for any vector x, where $x = (x_1, x_2, \ldots, x_n)^t \in \mathbb{C}^n$,

$$x^*Ax = \sum_{i,j} \bar{x}_i x_j a_{ij} = \sum_{i,j} \bar{x}_i x_j \langle u_j, u_i\rangle = \Big\langle \sum_j x_j u_j, \sum_i x_i u_i \Big\rangle \geq 0.$$

So $A \geq 0$. Conversely, if $A \geq 0$, we can write $A = B^*B$ for some n-square matrix B. (See Chapter 4.) Then $a_{ij} = b_i^* b_j = \langle b_j, b_i\rangle$, where b_i is the i-th column of B. So A is a Gram matrix.

5.13 If $A \geq 0$, then it is easy to check that $\langle x, y\rangle = y^*Ax$ is an inner product. Conversely, let $e_1, e_2, \ldots, e_n$ be an orthonormal basis of $\mathbb{C}^n$ and let $A = (\langle e_j, e_i\rangle)$. Then $A \geq 0$ and $\langle x, y\rangle = y_0^*Ax_0$, where x_0 and y_0 are the coordinates of x and y relative to the basis, respectively.

5.14 Compute $\langle u+v, u+v\rangle - \langle u-v, u-v\rangle$ directly.

5.15 No. Take $x = (1, 0)$ and $y = (1, 1)$.

5.16 By the Cauchy-Schwarz inequality, if x is a unit vector,

$$\langle x, y\rangle\langle y, x\rangle \leq \langle y, y\rangle.$$

Replace y by $\mathcal{A}x$,

$$\langle \mathcal{A}x, x\rangle\langle x, \mathcal{A}x\rangle \leq \langle \mathcal{A}x, \mathcal{A}x\rangle.$$

When $A \in M_n(\mathbb{C})$ and $x \in \mathbb{C}^n$ with $x^*x = 1$, $\langle Ax, x\rangle = x^*Ax$, so

$$x^*Axx^*A^*x \leq x^*A^*Ax.$$

The equality holds if and only if Ax and x are linearly dependent. Since $\|x\| = 1$, so $Ax = \lambda x$ for some λ, i.e., x is an eigenvector.

5.17 For the inequality, it is sufficient to notice that for all x

$$x^*(I + A)^{-1}(I - A)(I - A)(I + A)^{-1}x \leq x^*x$$

if and only if

$$(I + A)^{-1}(I - A)(I - A)(I + A)^{-1} \leq I,$$

which is true if and only if

$$(I - A)^2 \leq (I + A)^2.$$

This is obviously true.

For the equality case, we first show that $x \in \operatorname{Ker} A$ if and only if $(I-A)(I+A)^{-1}x = x$. If $x \in \operatorname{Ker} A$, then $Ax = 0$. Thus $(I-A)x = x$ and $(I + A)x = x$. Since $I + A$ is nonsingular, $x = (I + A)^{-1}x$. Thus $(I-A)(I+A)^{-1}x = (I-A)x = x$. Conversely, if $(I-A)(I+A)^{-1}x = x$, then since $I - A$ and $(I + A)^{-1}$ commute (because $I - A$ and $I + A$ commute), we have $(I + A)^{-1}(I - A)x = x$ or $(I - A)x = (I + A)x$, which implies $Ax = 0$; that is, $x \in \operatorname{Ker} A$.

We now show that the norm equality holds if and only if $(I+A)^{-1}(I-A)x = x$. If $(I+A)^{-1}(I-A)x = x$, the norm equality holds obviously. Conversely, if the norm equality holds, then x is an eigenvector of $M = (I-A)(I+A)^{-1}$. Let $Mx = \lambda x$. Since A is positive semidefinite, the eigenvalues of M should all have the form $(1-\mu)/(1+\mu)$, where μ is an eigenvalue of A. So $\lambda \neq 1$. Since $I - A$ and $(I+A)^{-1}$ commute, we have $(I+A)^{-1}(I-A)x = \lambda x$. It follows that $(1-\lambda)Ax = (\lambda-1)x$ and thus $Ax = -x$. This says -1 is an eigenvalue of A. A contradiction to A being positive semidefinite.

5.18 (a) Suppose that v_1, v_2, and v_3 are linearly dependent. It may be assumed that v_1, v_2, and v_3 are unit vectors and that

$$v_3 = \lambda_1 v_1 + \lambda_2 v_2.$$

Then

$$\langle v_1, v_3 \rangle = \lambda_1 + \lambda_2 \langle v_1, v_2 \rangle < 0 \quad \Rightarrow \quad \lambda_1 \langle v_1, v_2 \rangle + \lambda_2 \langle v_1, v_2 \rangle^2 > 0$$

and

$$\langle v_2, v_3 \rangle = \lambda_1 \langle v_1, v_2 \rangle + \lambda_2 < 0.$$

By subtracting

$$\lambda_2 (\langle v_1, v_2 \rangle^2 - 1) > 0 \quad \Rightarrow \quad \lambda_2 < 0 \quad \Rightarrow \quad \lambda_1 < 0.$$

Now compute

$$\langle v_4, v_3 \rangle = \lambda_1 \langle v_4, v_1 \rangle + \lambda_2 \langle v_4, v_3 \rangle > 0.$$

(b) No, since the dimension of the xy-plane is 2. However, it is possible for three vectors in the xy-plane to have pairwise negative products: $v_1 = (1, 0)$, $v_2 = (-1/2, \sqrt{3}/2)$, $v_3 = (-1/2, -\sqrt{3}/2)$.

(c) v_1, v_2, v_3, v_4 can linearly dependent or independent.

(d) $3, -\frac{3}{2}$. The maximum is attained when $u = v = w$, and the minimum is attained when the angles between any two of them are equal to $\frac{2\pi}{3}$.

5.19 (a) No.

(b) $\sqrt{\frac{3}{2}}x$, $\sqrt{\frac{5}{2}}x^2$.

(c) $\sqrt{\frac{3}{2}}x$, $\sqrt{\frac{5}{2}}x^2$, $\frac{3}{2\sqrt{2}} - \frac{5}{2\sqrt{2}}x^2$.

(d) It suffices to show that the basis vectors are orthogonal. For instance, by integration $\langle x, x^3 - \frac{3}{5}x \rangle = \int_{-1}^{1} (x^4 - \frac{3}{5}x^2) dx = 0$.

(e) It is sufficient to show that 1, x, $x^2 - \frac{1}{3}$, $x^3 - \frac{3}{5}x$ are linearly independent. Then it follows that $V_1 \cap V_2 = \{0\}$.

(f) Yes.

(g) Take $v = 1$ and $w = x$.

(h) No.

5.20 Let

$$x = \lambda_1 v_1 + \lambda_2 v_2 + \cdots + \lambda_n v_n.$$

Taking inner product of both sides with v_i results in

$$\lambda_i = \langle x, v_i \rangle, \quad i = 1, 2, \ldots, n.$$

By a direct computation,

$$\langle x, x \rangle = |\lambda_1|^2 + |\lambda_2|^2 + \cdots + |\lambda_n|^2.$$

5.21 (a) By definition.

(b) Note that $\operatorname{tr}(A^*A) = \langle A, A \rangle$.

(c) Use the Cauchy-Schwarz inequality.

(d) Note that $\operatorname{tr}(ABB^*A^*) = \operatorname{tr}(A^*ABB^*)$. Use the fact that

$$\operatorname{tr}(XY) \le \operatorname{tr} X \operatorname{tr} Y, \quad X \ge 0,\ Y \ge 0,$$

which can be shown by first assuming that X is diagonal.

(e) Since $\|A\|^2 = \langle A, A \rangle = \operatorname{tr}(A^*A)$ and $A^*A - AA^*$ is Hermitian,

$$\begin{aligned} \|A^*A - AA^*\|^2 &= \operatorname{tr}(A^*A - AA^*)^2 \\ &= 2\operatorname{tr}(A^*AA^*A) - 2\operatorname{tr}(A^*AAA^*) \\ &= 2\|A^*A\|^2 - 2\|A^2\|^2 \\ &\le 2\|A^*A\|^2 \\ &\le 2\|A\|^4. \end{aligned}$$

(f) Take $X = A^*$, then use (b). Or take $X = E_{ij}$, where E_{ij} is the matrix with 1 in the position (i, j) and 0 elsewhere.

(g) It is easy to show that W is a subspace. Note that the dimension of $M_n(\mathbb{C})$ is n^2. Since $\operatorname{tr} A = a_{11} + a_{22} + \cdots + a_{nn} = 0$, we have $\dim W = n^2 - 1$. Thus $\dim W^\perp = 1$. The identity matrix I is a basis for $W^\perp$ and $W^\perp$ only consists of scalar matrices. Alternatively, one may assume that A is upper-triangular, then show that A is a scalar matrix.

(h) Take $X = xx^*$ where x is a column vector.

No. Yes. No.

5.22 $u = (1, 0)$ is a unit vector, while $v = (1, -1)$ is not. They are not mutually orthogonal, since $\langle u, v \rangle = 1$.

(a) $u^\perp$ is the y-axis, $v^\perp$ is the line $y = x$.

(b) $u^\perp \cap v^\perp = \{0\}$.

(c) $\{u, v\}^\perp = \{0\}$.

(d) $(\text{Span}\{u, v\})^\perp = \{0\}$.

(e) $\text{Span}\{u^\perp, v^\perp\} = \mathbb{R}^2$.

5.23 $\begin{pmatrix} a & a \\ -b & -b \end{pmatrix}$, $a, b \in \mathbb{C}$.

5.24 (a) $x^*(A + cI)x = x^*Ax + c$.

(b) $x^*(cA)x = c(x^*Ax)$.

(c) Let e_i be the column vector with the i-th component 1 and 0 elsewhere. Then $e_i^*Ae_i = a_{ii} \in F(A)$.

(d) Let $Av = \lambda v$, where v is a unit vector. Then

$$v^*Av = v^*(\lambda v) = \lambda \in F(A).$$

(e) No, in general.

(f) A closed interval on the x-axis.

(g) A closed interval on the nonnegative part of the x-axis.

(h) The fields of values are, respectively, [0, 1]; the closed disc with center 0 and radius $\frac{1}{2}$; the closed elliptical disc with foci at 0 and 1, minor axis 1 and major axis $\sqrt{2}$; the closed line segment joining (1, 0) and (1, 1); and the triangle (including interior) with the vertices (0, 0), (1, 0), and (1, 1).

5.25 (a) By a direct computation.

(b) Suppose $\|x + y\| = \|x\| + \|y\|$. On one hand, for $s, t \geq 0$,

$$\|sx + ty\| \leq s\|x\| + t\|y\|.$$

On the other hand, assuming $t \geq s$ ($s > t$ is dealt similarly),

$$\begin{aligned} \|sx + ty\| &= \|t(x + y) - (t - s)x\| \\ &\geq |\, t\|x + y\| - (t - s)\|x\| \,| \\ &= s\|x\| + t\|y\|. \end{aligned}$$

Thus $\|sx + ty\| = s\|x\| + t\|y\|$. The other direction is obvious. This result can also be shown by examining the proof of the triangle inequality for inner product spaces.

(c) If x and y are orthogonal, a simple computation gives

$$\|x+y\|^2 = \|x\|^2 + \|y\|^2 + \langle x, y\rangle + \langle y, x\rangle = \|x\|^2 + \|y\|^2.$$

(d) By a direct computation. This is the Pythagorean Theorem.

(e) The converse of (d) is true over $\mathbb{R}$; false over $\mathbb{C}$: $x = 1$, $y = i$.

5.26 Necessity: Let $v \in W$ and $u = v + v'$, where $v' \in W^\perp$. Then $u - v \in W^\perp$. Thus for every $w \in W$, $v - w \in W$, $\langle u-v,\ v-w\rangle = 0$. Therefore,

$$\|u-w\|^2 = \|(u-v)+(v-w)\|^2 = \|v-v\|^2 + \|v-w\|^2$$

and

$$\|u-v\| \leq \|u-w\|.$$

Sufficiency: Assume that v_1 is the projection of u onto W and $v_1 \neq v$. Let $u = v_1 + v_2$, where $v_1 \in W$, $v_2 \in W^\perp$. Then $(u - v_1) \perp (v_1 - v)$,

$$\begin{aligned} \|u-v\|^2 &= \|(u-v_1)+(v_1-v)\|^2 \\ &= \|u-v_1\|^2 + \|v_1-v\|^2 \\ &> \|u-v_1\|^2, \end{aligned}$$

a contradiction.

5.27 First, we claim $W \cap W^\perp = \{0\}$. If x is contained in W and $W^\perp$, then $\langle x, x\rangle = 0$, so $x = 0$. Thus $W + W^\perp$ is a direct sum. So $\dim W + \dim W^\perp \leq \dim V$. Now we show that the direct sum is indeed equal to V. If $W = \{0\}$ or V, we have nothing to show. Suppose $\dim W = s$, $0 < s < n = \dim V$. Let $\alpha_1, \alpha_2, \ldots, \alpha_s$ be an orthogonal basis for W. $W^\perp$ cannot be $\{0\}$; otherwise $W = V$. Let $0 \neq \beta_1 \in W^\perp$. Then $\alpha_1, \alpha_2, \ldots, \alpha_s, \beta_1$ are pairwise orthogonal and linearly independent. Let $W_1 = \operatorname{Span}\{\alpha_1, \alpha_2, \ldots, \alpha_s, \beta_1\}$. If $W_1 = V$, then we are done. Otherwise, choose $0 \neq \beta_2 \in W_1^\perp$. Inductively, we have a set of nonzero vectors $\beta_1, \beta_2, \ldots, \beta_t$ such that they all lie in $W^\perp$ and are mutually orthogonal, $s + t = n$. So $V = W \oplus \operatorname{Span}\{\beta_1, \beta_2, \ldots, \beta_t\}$. We show that $W^\perp = \operatorname{Span}\{\beta_1, \beta_2, \ldots, \beta_t\}$. Since $\beta_1, \beta_2, \ldots, \beta_t \in W^\perp$, we have $\operatorname{Span}\{\beta_1, \beta_2, \ldots, \beta_t\} \subseteq W^\perp$. However, $\dim W^\perp \leq n - s = t$, $\operatorname{Span}\{\beta_1, \beta_2, \ldots, \beta_t\} = W^\perp$ and the desired conclusion follows.

5.28 (a) False.

(b) False.

(c) True.

5.29 (a) True.

(b) True.

(c) False, unless S is a subspace.

(d) True.

(e) True.

(f) True.

(g) True.

(h) True.

5.30 We show (a). (b) is similar. Since $W_1 \subseteq W_1+W_2$, $(W_1+W_2)^\perp \subseteq W_1^\perp$. Likewise, $(W_1 + W_2)^\perp \subseteq W_2^\perp$. So $(W_1 + W_2)^\perp \subseteq W_1^\perp \cap W_2^\perp$. Now suppose $u \in W_1^\perp \cap W_2^\perp$. Then $\langle u, w_1 \rangle = 0$ for all $w_1 \in W_1$ and $\langle u, w_2 \rangle = 0$ for all $w_2 \in W_1$. Thus $\langle u, w \rangle = 0$ for all $w \in W_1 + W_2$.

5.31 Since $p + \dim S^\perp = \dim V = n$, $\dim S^\perp = n - p < q$. The vectors v_i are all contained in $S^\perp$, they must be linearly dependent.

5.32 Notice that

$$\begin{pmatrix} x^* & 0 \\ 0 & y^* \end{pmatrix} \begin{pmatrix} A & B^* \\ B & C \end{pmatrix} \begin{pmatrix} x & 0 \\ 0 & y \end{pmatrix} = \begin{pmatrix} x^*Ax & x^*B^*y \\ y^*Bx & y^*Cy \end{pmatrix} \geq 0.$$

Take the determinant. For the particular cases, observe that

$$\begin{pmatrix} A & A \\ A & A \end{pmatrix} \geq 0 \quad \text{and} \quad \begin{pmatrix} A & I \\ I & A^{-1} \end{pmatrix} \geq 0.$$

5.33 Let x be an eigenvector of $\mathcal{A}$ corresponding to λ. Then $\langle \mathcal{A}x, x \rangle = -\langle x, \mathcal{A}x \rangle$ implies that $\langle \lambda x, x \rangle = -\langle x, \lambda x \rangle$. Thus $\lambda \langle x, x \rangle = -\bar{\lambda} \langle x, x \rangle$. It follows that $\lambda + \bar{\lambda} = 0$ and $\lambda = 0$ if λ is real.

5.34 $S = \operatorname{Span}\{u_1, u_2\}$, where $u_1 = (-3, 0, 1, 0)$, $u_2 = (2, -1, 0, 1)$ and $S^\perp = \operatorname{Span}\{v_1, v_2\}$, where $v_1 = (1, 0, 3, 1)$, $v_2 = (0, 1, 0, 1)$.

5.35 Let $\dim V = n$, $\dim V_1 = s$, $\dim V_2 = t$. Then $\dim V_1^\perp = n - s \geq 1$. Let $V_3 = V_2 \cap V_1^\perp$. Then by the dimension identity, we have

$$\dim V_3 = \dim V_2 + \dim V_1^\perp - \dim(V_2 + V_1^\perp) \geq t + (n - s) - n = t - s > 0.$$

This implies that $V_3 = V_2 \cap V_1^\perp \neq \{0\}$. Thus for some $u \in V_2$, $u \neq 0$, $\langle x, u \rangle = 0$ for all $x \in V_1$.

5.36 Let λ be an eigenvalue of $\mathcal{A}$ and V_λ be the eigenspace of λ. Write $V = V_\lambda \oplus V_\lambda^\perp$. We show that $V_\lambda^\perp$ is invariant under $\mathcal{A}$. Let $x \in V_\lambda^\perp$, then $\langle x, y\rangle = 0$ for all $y \in V_\lambda$. Thus $\langle \mathcal{A}(x), y\rangle = \langle x, \mathcal{A}(y)\rangle = \langle x, \lambda y\rangle = \bar{\lambda}\langle x, y\rangle = 0$; that is, $\mathcal{A}(x) \in V_\lambda^\perp$. Now use induction on the dimension of the vector space.

5.37 Let $\lambda_1, \lambda_2, \ldots, \lambda_n$ be the eigenvalues of $\mathcal{A}$, and let $v_1, v_2, \ldots, v_n$ be the corresponding eigenvectors that form an orthonormal basis for V. Suppose further that the first m eigenvalues are positive and the rest are not. If $m < k$, then there exists a nonzero vector w such that

$$w \in W \cap \operatorname{Span}\{v_{m+1}, \ldots, v_n\}.$$

Write

$$w = c_{m+1}v_{m+1} + \cdots + c_n v_n.$$

Since $\{v_1, v_2, \ldots, v_n\}$ is an orthonormal set,

$$\langle \mathcal{A}w, w\rangle = \Big\langle \sum_{i=m+1}^{n} c_i\lambda_i v_i, \sum_{i=m+1}^{n} c_i v_i \Big\rangle = \sum_{i=m+1}^{n} |c_i|^2\lambda_i \le 0.$$

This is a contradiction. Therefore, $m \ge k$.

5.38 It is sufficient to show that

$$\mathcal{A} = 0 \Leftrightarrow \langle \mathcal{A}v, w\rangle = 0, \quad \text{for all } v \in V \text{ and } w \in W.$$

If $\mathcal{A}v \neq 0$ for some v, let $w = \mathcal{A}v$. Then $\langle \mathcal{A}v, w\rangle \neq 0$, a contradiction.

5.39 (a) If $\langle x, \mathcal{B}y\rangle = \langle \mathcal{A}x, y\rangle = \langle x, \mathcal{A}^*y\rangle$ for all x and y, $\mathcal{B} = \mathcal{A}^*$.

(b) Note that

$$\langle \mathcal{A}x, y\rangle = \langle x, \mathcal{A}^*y\rangle = \overline{\langle \mathcal{A}^*y, x\rangle} = \overline{\langle y, (\mathcal{A}^*)^*x\rangle} = \langle (\mathcal{A}^*)^*x, y\rangle.$$

(c) If $x \in \operatorname{Ker}\mathcal{A}^*$, then $\mathcal{A}^*x = 0$. For any $y \in V$,

$$\langle x, \mathcal{A}y\rangle = \langle \mathcal{A}^*x, y\rangle = 0.$$

Hence $\operatorname{Ker}\mathcal{A}^* \subseteq (\operatorname{Im}\mathcal{A})^\perp$. The other way is similarly shown.

(d) Similar to (c).

(e) By (c) and (d).

(f) This is because $\langle \mathcal{A}\alpha_i, \alpha_j\rangle = \langle \alpha_i, \mathcal{A}^*\alpha_j\rangle = \overline{\langle \mathcal{A}^*\alpha_j, \alpha_i\rangle}$.

5.40 It is an inner product if and only if $\mathcal{A}$ is invertible.

5.41 Compute $\langle \mathcal{A}(x+y), \mathcal{A}(x+y)\rangle$.

5.42 (b) is equivalent to $\langle \mathcal{A}(u), \mathcal{A}(u)\rangle = \langle u, u\rangle$. So (a) implies (b). For (b)$\Rightarrow$(c), compute $\langle \mathcal{A}(\alpha_i+\alpha_j), \mathcal{A}(\alpha_i+\alpha_j)\rangle$ to get $\langle \mathcal{A}(\alpha_i), \mathcal{A}(\alpha_j)\rangle = 1$ if $i=j$, 0 otherwise. To show that (c)$\Rightarrow$(d), let A be the matrix representation of $\mathcal{A}$ under the basis $\alpha_1, \ldots, \alpha_n$. Let a_i be the i-th column of A. Then $\langle a_i, a_j\rangle = \langle \mathcal{A}(\alpha_i), \mathcal{A}(\alpha_j)\rangle = 1$ if $i = j$, 0 otherwise. Thus A is orthogonal. For (d)$\Rightarrow$(a), write u and v as linear combinations of $\alpha_1, \alpha_2, \ldots, \alpha_n$, say, x and y, respectively. Then $\langle u, v\rangle = x^t y$ and $\langle \mathcal{A}(u), \mathcal{A}(v)\rangle = (Ax)^t(Ay) = x^t A^t Ay = x^t y$. Condition (c) cannot be replaced by (c′): Take $\mathcal{A}(e_1) = 2e_1$ and $\mathcal{A}(e_2) = 2e_2$ for $\mathbb{R}^2$.

5.43 $\mathcal{A}$ is not necessarily orthogonal in general. Take $\mathbb{R}^2$ with $e_1 = (1,0)$, $e_2 = (0,1)$ and define $\mathcal{A}e_1 = e_1$ and $\mathcal{A}e_2 = e_1$; that is, $\mathcal{A}(e_1, e_2) = (e_1, e_2)A$, where $A = \left(\begin{smallmatrix} 1 & 1 \\ 0 & 0 \end{smallmatrix}\right)$. Then $\mathcal{A}$ is a linear transformation satisfying $\langle \mathcal{A}e_i, \mathcal{A}e_i\rangle = \langle e_1, e_1\rangle = 1, i = 1, 2$. But $\mathcal{A}$ is not orthogonal, since $\langle e_1, e_2\rangle = 0$, but $\langle \mathcal{A}e_1, \mathcal{A}e_2\rangle = \langle e_1, e_1\rangle = 1 \neq 0$.

5.44 Take $V = \mathbb{R}^2$ and define $\mathcal{L}(x,y) = \left(\sqrt{(x^2+y^2)/2}, \sqrt{(x^2+y^2)/2}\,\right)$ and $\mathcal{D}(x,y) = (x+1, y+1)$. Then $\mathcal{L}$ preserves length, $\mathcal{D}$ preserves distance, but neither $\mathcal{L}$ nor $\mathcal{D}$ is linear.

5.45 No, in general. Yes, if $\alpha_1, \ldots, \alpha_n$ are linearly independent. For the orthogonal case, the sufficiency is obvious by Problem 5.42. For the necessity, let, without loss of generality, $\{\alpha_1, \ldots, \alpha_t\}$ be a basis for $\operatorname{Span}\{\alpha_1, \ldots, \alpha_n\}$. Then it can be shown that $\{\beta_1, \ldots, \beta_t\}$ is a basis for $\operatorname{Span}\{\beta_1, \ldots, \beta_n\}$. Now let $\{u_1, \ldots, u_{n-t}\}$ and $\{v_1, \ldots, v_{n-t}\}$ be orthonormal bases, respectively, for

$$(\operatorname{Span}\{\alpha_1, \ldots, \alpha_t\})^\perp \quad \text{and} \quad (\operatorname{Span}\{\beta_1, \ldots, \beta_t\})^\perp.$$

Then

$$\{\alpha_1, \ldots, \alpha_t, u_1, \ldots, u_{n-t}\} \quad \text{and} \quad \{\beta_1, \ldots, \beta_t, v_1, \ldots, v_{n-t}\}$$

are two bases for V. Now suppose $x \in V$. If

$$x = \sum_{i=1}^{t} x_i \alpha_i + \sum_{i=1}^{n-t} y_i u_i,$$

let

$$\mathcal{A}x = \sum_{i=1}^{t} x_i \beta_i + \sum_{i=1}^{n-t} y_i v_i.$$

Then $\mathcal{A}$ is an orthogonal linear transformation.

5.46 Show first that $\langle \mathcal{A}(x), \mathcal{A}(y)\rangle = \langle \mathcal{B}(x), \mathcal{B}(y)\rangle$ for all x and y by considering $\langle \mathcal{A}(x+y), \mathcal{A}(x+y)\rangle$. Take a basis $\alpha_1, \ldots, \alpha_n$ for V. For the sets $\{\mathcal{A}(\alpha_1), \ldots, \mathcal{A}(\alpha_n)\}$ and $\{\mathcal{B}(\alpha_1), \ldots, \mathcal{B}(\alpha_n)\}$, by Problem 5.45, there exists an orthogonal transformation $\mathcal{C}$ such that $\mathcal{C}(\mathcal{B}(\alpha_i)) = \mathcal{A}(\alpha_i)$. It follows that $\mathcal{A} = \mathcal{C}\mathcal{B}$.

5.47 (a) By definition

$$\begin{aligned} \mathcal{L}_v(ax+by) &= \langle ax+by, v\rangle \\ &= a\langle x, v\rangle + b\langle y, v\rangle \\ &= a\mathcal{L}_v(x) + b\mathcal{L}_v(y). \end{aligned}$$

(b) Notice that $\mathcal{L}(ax+by) = \mathcal{L}_{ax+by}$ and that

$$\begin{aligned} \mathcal{L}_{ax+by}(u) &= \langle u, ax+by\rangle \\ &= a\langle u, x\rangle + b\langle u, y\rangle \\ &= a\mathcal{L}_x(u) + b\mathcal{L}_y(u) \\ &= (a\mathcal{L}_x + b\mathcal{L}_y)(u). \end{aligned}$$

Thus $\mathcal{L}$ is linear.

(c) If $\mathcal{L}(v_1) = \mathcal{L}(v_2)$, then $\mathcal{L}_{v_1} = \mathcal{L}_{v_2}$ and

$$\langle u, v_1\rangle = \langle u, v_2\rangle, \quad \text{for every } u \in V.$$

Thus $v_1 = v_2$ and $\mathcal{L}$ is one-to-one.

To show that $\mathcal{L}$ is onto, suppose that $\{e_1, e_2, \ldots, e_n\}$ is an orthonormal basis for V. If $f \in V^*$, let

$$v = f(e_1)e_1 + \cdots + f(e_n)e_n.$$

Then $\mathcal{L}_v = f$.

(d) For orthonormal basis $\{e_1, e_2, \ldots, e_n\}$ of V, define $f_i \in V^*$ by

$$f_i(e_j) = \begin{cases} 1 & \text{if } i = j, \\ 0 & \text{if } i \neq j. \end{cases}$$

Then $\{f_1, f_2, \ldots, f_n\}$ is a basis for V^*.

5.48 First show $W_1 \cap W_2 = \{0\}$. If $\alpha \in W_1 \cap W_2$. Then $\alpha = \mathcal{T}(\alpha)$ and $\alpha = \beta - \mathcal{T}(\beta)$ for some $\beta \in V$. Compute $\langle \alpha, \alpha\rangle = \langle \alpha, \beta - \mathcal{T}(\beta)\rangle = \langle \alpha, \beta\rangle - \langle \alpha, \mathcal{T}(\beta)\rangle = \langle \alpha, \beta\rangle - \langle \mathcal{T}(\alpha), \mathcal{T}(\beta)\rangle = \langle \alpha, \beta\rangle - \langle \alpha, \beta\rangle = 0$. So $\alpha = 0$. Since $x = \mathcal{T}(x) + (x - \mathcal{T}(x))$ for all $x \in V$, $V = W_1 \oplus W_2$.

5.49 (a) Extend u to an orthonormal basis $\{u, u_2, \dots, u_n\}$. Then

$$\mathcal{A}(u) = -u, \quad \mathcal{A}(u_i) = u_i, \quad i = 2, \dots, n.$$

(b) The matrix of $\mathcal{A}$ under the above basis is $A = \begin{pmatrix} -1 & 0 \\ 0 & I_{n-1} \end{pmatrix}$. Thus $|A| = -1$. A matrix representation of $\mathcal{A}$ under a different basis is similar to A. They have the same determinant.

(c) Suppose the coordinate vectors of x and $\mathcal{A}(x)$ on the orthonormal basis are y_0 and z_0, respectively. Let the coordinate vector of u on the basis be v. Then

$$z_0 = y_0 - 2(v^t y_0)v = Iy_0 - 2(vv^t)y_0 = (I - 2vv^t)y_0.$$

Thus the matrix of $\mathcal{A}$ is of the form $I - 2vv^t$.

(d) Apply $\mathcal{A}$ to $x = ku + y$.

(e) Denote by V_1 the eigenspace of 1 and suppose that

$$\{u_1, u_2, \dots, u_{n-1}\}$$

is a basis for V_1. Let $\alpha \neq 0$ be an eigenvalue of $\mathcal{B}$ having unit eigenvector $u_n \in V_1^{\perp}$. Then $u_1, u_2, \dots, u_n$ form a basis for V. Considering the matrix representation of $\mathcal{B}$ on this basis, one has $\alpha = -1$. Thus

$$\mathcal{B}u_i = u_i, \quad i = 1, 2, \dots, n-1, \quad \mathcal{B}u_n = -u_n.$$

Take $w = u_n$, then $\mathcal{B}(x) = x - 2\langle x, w\rangle w$.

5.50 (a) For $v \in V$, write $v = w + w^{\perp}$, where $w \in W$ and $w^{\perp} \in W^{\perp}$. Define $\mathcal{P}(v) = w$.

(b) $\mathcal{P}^2(v) = \mathcal{P}(w) = w = \mathcal{P}(v)$.

(c) The decomposition $v = w + w^{\perp}$ is uniquely determined by W.

(d) Since W is a nontrivial subspace, there exists a subspace W' such that (see Problem 1.48)

$$V = W \oplus W' \quad \text{and} \quad W' \neq W^{\perp}.$$

Similarly, define $\mathcal{P}'$ as $\mathcal{P}$.

(e) $\langle \mathcal{P}(v), v\rangle = \langle w, w + w^{\perp}\rangle = \langle w, w\rangle \geq 0$.

(f) $\|\mathcal{P}(v)\| = \|w\| \leq \|w + w^{\perp}\| = \|v\|$.

(g) $(\mathcal{I} - \mathcal{P})(w^\perp) = w^\perp - \mathcal{P}w^\perp = w^\perp$.

(h) $\langle \mathcal{P}(v), (\mathcal{I} - \mathcal{P})(v) \rangle = 0$.

5.51 (a) For $v \in V$,

$$v = \mathcal{I}v = \mathcal{P}_1 v + \cdots + \mathcal{P}_m v.$$

Thus

$$V = \operatorname{Im} \mathcal{P}_1 + \cdots + \operatorname{Im} \mathcal{P}_m.$$

To see it is a direct sum, we show that

$$\dim V = \dim(\operatorname{Im} \mathcal{P}_1) + \cdots + \dim(\operatorname{Im} \mathcal{P}_m).$$

Take a basis for V and suppose that the matrix representation for $\mathcal{P}_i$ is P_i, $i = 1, \ldots, m$. By Problem 3.104,

$$\dim(\operatorname{Im} \mathcal{P}_i) = \operatorname{tr} P_i = r(P_i), \quad i = 1, \ldots, m.$$

It follows that

$$\begin{aligned} \dim V &= n \\ &= \operatorname{tr} I_n \\ &= \operatorname{tr} P_1 + \cdots + \operatorname{tr} P_m \\ &= \dim(\operatorname{Im} \mathcal{P}_1) + \cdots + \dim(\operatorname{Im} \mathcal{P}_m). \end{aligned}$$

To see $\mathcal{P}_i \mathcal{P}_j = 0$ for distinct i and j, let $v \in V$. Then

$$\mathcal{P}_j v = \Big(\sum_{i=1}^{n} \mathcal{P}_i \Big) \mathcal{P}_j v = \sum_{i=1}^{n} (\mathcal{P}_i \mathcal{P}_j) v.$$

Note that

$$\mathcal{P}_j v \in \operatorname{Im} \mathcal{P}_j, \quad \mathcal{P}_i \mathcal{P}_j v \in \operatorname{Im} \mathcal{P}_i$$

and that V is a direct sum of $\operatorname{Im} \mathcal{P}_j$'s. It follows that

$$(\mathcal{P}_i \mathcal{P}_j) v = 0, \quad i \neq j.$$

(b) Choose a basis for each $\operatorname{Im} \mathcal{P}_i$, $i = 1, \ldots, m$, and put all these vectors together to form an orthonormal basis for V by defining the inner product of any two distinct vectors in the basis to be 0, and the inner product of any vector with itself to be 1.

(c) Obviously

$$\operatorname{Im}\mathcal{P}_1 + \cdots + \operatorname{Im}\mathcal{P}_m + \cap_{i=1}^m \operatorname{Ker}\mathcal{P}_i \subseteq V.$$

Let

$$\mathcal{T} = \mathcal{I} - \mathcal{P}_1 - \cdots - \mathcal{P}_m.$$

Then for $x \in V$,

$$\mathcal{T}x = x - \mathcal{P}_1 x - \cdots - \mathcal{P}_m x.$$

Thus for each i,

$$\begin{aligned} \mathcal{P}_i(\mathcal{T}x) &= \mathcal{P}_i(x - \mathcal{P}_1 x - \cdots - \mathcal{P}_m x) \\ &= \mathcal{P}_i x - \mathcal{P}_i^2 x = 0. \end{aligned}$$

So

$$\mathcal{T}x \in \cap_{i=1}^m \operatorname{Ker}\mathcal{P}_i.$$

However,

$$x = \mathcal{P}_1 x + \cdots + \mathcal{P}_m x + \mathcal{T}x,$$

hence

$$V = \operatorname{Im}\mathcal{P}_1 + \cdots + \operatorname{Im}\mathcal{P}_m + \cap_{i=1}^m \operatorname{Ker}\mathcal{P}_i.$$

To show it is a direct sum, let

$$x_1 + \cdots + x_m + y = 0,$$

where

$$x_i \in \operatorname{Im} P_i, \quad y \in \cap_{i=1}^m \operatorname{Ker}\mathcal{P}_i.$$

Notice that $\mathcal{P}_i x_i = x_i$ and that $\mathcal{P}_i\mathcal{P}_j = 0$. Applying $\mathcal{P}_i$ to both sides of the above identity yields $x_i = 0$, $i = 1, 2, \ldots, m$, $y = 0$.

Notation

$\mathbb{R}$	real numbers
$\mathbb{R}^+$	positive numbers
$\mathbb{C}$	complex numbers
$\mathbb{Q}$	rational numbers
$\mathbb{F}$	a field
$\mathbb{R}^n$	column vectors with n real components
$\mathbb{C}^n$	column vectors with n complex components
$\mathbb{P}_n[x]$	real polynomials with degree less than n
$\mathbb{P}[x]$	real polynomials with any degree
$\|c\|$	absolute value of complex number c
$\bar{c}$	conjugate of complex number c
$\operatorname{Re} c$	real part of complex number c
$\mathcal{C}[a, b]$	real-valued continuous functions on $[a, b]$
$\mathcal{C}(\mathbb{R})$	real-valued continuous functions
$\mathcal{C}_\infty(\mathbb{R})$	real-valued functions of derivatives of all orders
$P \Rightarrow Q$	If P then Q
$P \Leftrightarrow Q$	P if and only if Q
V, W	vector spaces
W_1, W_2	subspaces
$\operatorname{Span} S$	the vector space generated by the vectors in S
$e_1, \ldots, e_n$	standard basis for $\mathbb{R}^n$ or $\mathbb{C}^n$
$M_{m \times n}(\mathbb{F})$	$m \times n$ matrices with entries in $\mathbb{F}$
$M_n(\mathbb{F})$	$n \times n$ matrices with entries in $\mathbb{F}$
$H_n(\mathbb{F})$	$n \times n$ Hermitian matrices with entries in $\mathbb{F}$
$S_n(\mathbb{F})$	$n \times n$ Skew-Hermitian matrices with entries in $\mathbb{F}$
$\frac{df}{dt}, \; '$	derivative of f with respect to t
y''	second derivative of y
I_n, I	$n \times n$ identity matrix
$\dim V$	dimension of vector space V
$W_1 + W_2$	sum of W_1 and W_2
$W_1 \oplus W_2$	direct sum of W_1 and W_2
$A, B, \ldots$	matrices
$\mathcal{A}, \mathcal{B}, \ldots$	linear transformations
E_{ij}	square matrix with the (i, j)-entry 1 and 0 elsewhere
$\|A\|$	determinant of matrix A

$\det A$	determinant of matrix A
$r(A)$	rank of matrix A
$\operatorname{tr} A$	trace of matrix A
A^t	transpose of matrix A
$\bar{A}$	conjugate of matrix A
A^*	conjugate transpose of matrix A
A^{-1}	inverse of matrix A
$\operatorname{adj}(A)$	adjoint matrix of matrix A
$\operatorname{diag}(\lambda_1, \ldots, \lambda_n)$	diagonal matrix with $\lambda_1, \ldots, \lambda_n$ on the main diagonal
$\operatorname{Ker} A$	kernel or null space of A, i.e., $\operatorname{Ker} A = \{x \mid Ax = 0\}$
$\operatorname{Im} A$	image or range of A, i.e., $\operatorname{Im} A = \{Ax\}$
$F(A)$	field of values of A, i.e., $\{x^*Ax \mid \|x\| = 1\}$
$\|\lambda I - A\|$	characteristic polynomial of A
$\lambda_{\max}(A)$	largest eigenvalue of matrix A
$\sigma_{\max}(A)$	largest singular value of matrix A
$\|x\|$	norm or length of vector x, i.e., $\sqrt{x^*x}$ or $\sqrt{\langle x, x\rangle}$
$A \geq 0$	A is a positive semidefinite matrix
$A \geq B$	$A - B \geq 0$
$A > 0$	A is a positive definite matrix
$A^{\frac{1}{2}}$	square root of positive semidefinite matrix A
$\mathrm{m}(A)$	the modulus of A, i.e., $\mathrm{m}(A) = (A^*A)^{\frac{1}{2}}$
$[A, B]$	commutator $AB - BA$
$A \circ B$	entrywise product of A and B, i.e., $A \circ B = (a_{ij}b_{ij})$
$\mathcal{A}^*$	adjoint of linear transformation $\mathcal{A}$
$W^{\perp}$	subspace of the vectors orthogonal to W
V^*	dual space
$\mathcal{P}$	orthogonal projection
$\begin{vmatrix} A & B \\ C & D \end{vmatrix}$	determinant of the block matrix $\begin{pmatrix} A & B \\ C & D \end{pmatrix}$

Main References

Carlson D., C. Johnson, D. Lay, and A. Porter. *Linear Algebra Gems.* Washington DC: Mathematical Association of America, 2002.

Horn R. A., and C. R. Johnson. *Matrix Analysis.* Cambridge: Cambridge University Press, 1985.

Horn R. A., and C. R. Johnson. *Topics in Matrix Analysis.* Cambridge: Cambridge University Press, 1991.

Marcus M., and H. Minc. *A Survey of Matrix Theory and Matrix Inequalities.* New York: Dover, 1992.

Marshall A. W., and I. Olkin. *Inequalities: Theory of Majorization and Its Applications.* New York: Academic Press, 1979.

Ou W.-Y., C.-X. Li, and P. Zhang. *Graduate Entrance Exams in Math with Solutions.* Changchun: Jilin University Press, 1998 (in Chinese).

Qian J.-L. *Selected Problems in Higher Algebra.* Beijing: Central University of Nationalities Press, 2002 (in Chinese).

Shi M.-R.. *600 Linear Algebra Problems with Solutions.* Beijing: Beijing Press of Science and Technology, 1985 (in Chinese).

Wang B.-Y. *Introduction to Majorization and Matrix Inequalities.* Beijing: Beijing Normal University Press, 1990 (in Chinese).

Zhang F. *Matrix Theory: Basic Results and Techniques.* New York: Springer, 1999.

Index